SAVING OUR STURGEON

Protecting Wisconsin's Ancient Fish

Rebecca Hogue Wojahn

WISCONSIN HISTORICAL SOCIETY PRESS

Published by the Wisconsin Historical Society Press
Publishers since 1855

The Wisconsin Historical Society helps people connect to the past by collecting, preserving, and sharing stories. Founded in 1846, the Society is one of the nation's finest historical institutions.
Join the Wisconsin Historical Society: wisconsinhistory.org/membership

Publication of this book was made possible in part by a grant from the D.C. Everest fellowship fund.

Cover illustration by Camille Davis/Cricket Design Works

Printed in Wisconsin, USA
Designed by Cricket Design Works and John Ferguson

29 28 27 26 25 1 2 3 4 5

Library of Congress Cataloging-in-Publication Data available.

∞ The paper used in this publication meets the minimum requirements of the American National Standard for Information Sciences—Permanence of Paper for Printed Library Materials, ANSI Z39.48-1992.

For Wisconsin sturgeon enthusiasts—past, present, and future.

—RHW

Places in This Book

Contents

CHAPTER 1

LIVING DINOSAURS

"Clear the way!" the fish biologist calls. She and her coworker heave a net with a huge fish up the rocky bank of the Wolf River, outside of Shiocton, Wisconsin. The thrashing fish looks like something plucked from the dinosaur age. Its grayish brown skin is marked with armored bumps along its back and sides. Whiskers hang down from its snout. Its toothless, tubelike mouth extends beneath its head, gaping for water.

The biologists move quickly. Their chest-high wader boots rumble and squelch as they wrestle the netted fish through the crowd of people that has gathered on the shore to watch. They heft it toward some University of Wisconsin–Stevens Point student volunteers in baseball caps, fishing gloves, and their own waders. Water, sweat, and fish slime drench everyone's sweatshirts. But grins dance on their faces.

The fish is a lake sturgeon, one of the largest and oldest fish **species** on the planet. Each spring, when the river warms to about 53 degrees Fahrenheit, **mature** sturgeon return to the place on the river where they were born to lay and **fertilize** eggs. And every spring, sturgeon biologists are there to meet them. It's the one time of the year when these giant, solitary fish come together in large numbers, making it easier for the biologists to count and study them.

Wisconsin Department of Natural Resources (DNR) staff haul a lake sturgeon out of the water at Bamboo Bend on the Wolf River.

The sturgeon is swung onto a tarp at the station, and the college students get to work. As they untangle the sturgeon from the net, one student straddles its wide head and another attempts to pin its tail against a long measuring stick. "Female, sixty-two inches!" the student calls. The volunteer on clipboard duty jots this information down, along with the weight of the fish.

A biologist waves a scanner the size and shape of a clothes iron down the length of the sturgeon's twisting body. "No PIT," he says. He pulls a syringe from a plastic bin. With a quick jab, he places a new **PIT (Passive Integrated Transponder) tag** under the fish's skin. The PIT tag, which is the size of a grain of rice, will allow the scientists to record this sturgeon's movements, growth, and more if and when it is captured again. It's the same technology used for microchipping family pets.

The sturgeon rolls and flops on the tarp, whacking one of the students across the chest with her sharklike tail fin. The crowd around the station chuckles and snaps pictures. The students wrestle the sturgeon down a tarp-covered chute that leads back into the river. "Time for you to head back home," one of the volunteers says with a gentle shove.

The fish doesn't hesitate. She arcs and spins and belly flops into the current. As soon as she hits the water, she makes her way back upriver to the rocks where she will release her eggs. Meanwhile, the volunteers on

With crowds looking on, UW–Stevens Point student volunteers wrestle a lake sturgeon to measure its length. A third student (out of frame) records the data.

Student volunteers slide the sturgeon down a tarp back into the Wolf River.

the shore also look upriver. The biologists have already netted another sturgeon and are lugging it their way.

This scene is repeated, fish by fish, for multiple days and at multiple stations along Wisconsin's Fox, Peshtigo, and Wolf Rivers.

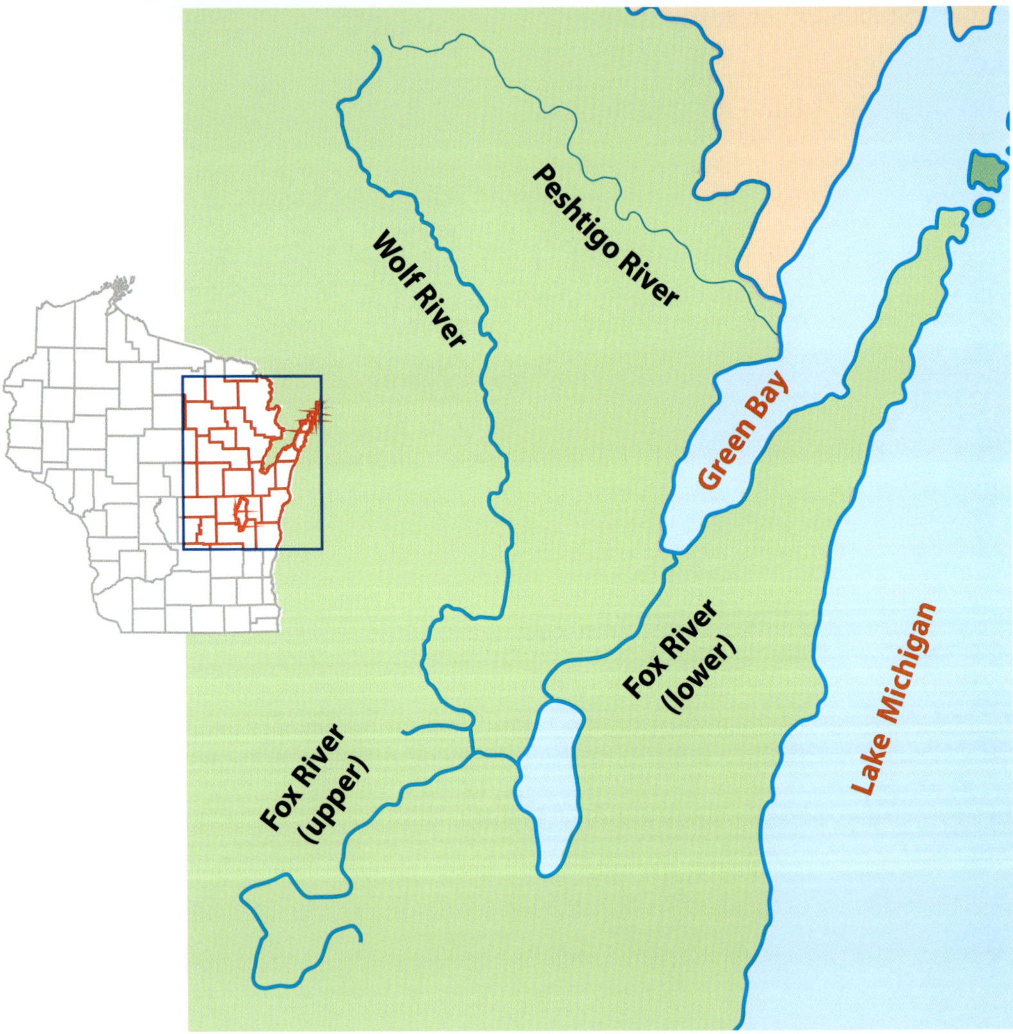

It's an annual event that's been happening for decades. This careful monitoring is one reason most of Wisconsin's lake sturgeon population is steady and healthy. In other parts of the United States and the world, the species' status is less stable.

It's not just the sturgeon that the biologists learn about with this monitoring routine. Margaret Stadig, the Wisconsin Department of Natural Resources (DNR) fisheries biologist who oversees lake sturgeon in Wisconsin, explains:

> [Sturgeon] are a great **indicator species** on how the **ecosystem** as a whole is doing. If an ecosystem is struggling, the likelihood is you won't see sturgeon populations in the system. If a system is fairly healthy, then you tend to see more sturgeon. It indicates your ecosystem as a whole is doing very well, particularly when you see little [as in, small] sturgeon consistently.

But it wasn't always this way for Wisconsin's water systems and its lake sturgeon. There was a time when lake sturgeon almost completely died out in Wisconsin waters. Humans were responsible for causing that near disaster. But humans were also responsible for fixing it.

This is the story of how, over the last 150 years, we saved an ancient fish in Wisconsin.

Threatened, Endangered, Extinct: What's Behind the Categories?

Many governments and organizations assign a **conservation** status to plant and animal species. For example, the US government has an Endangered Species List. Two US government departments are responsible for managing the list: the US Fish and Wildlife Service (USFWS) and the National Marine Fisheries Service (NMFS).

Some organizations track species all over the world, while others focus on a particular region. You may be familiar with some conservation status categories, such as "threatened," "endangered," and "extinct." These categories tell us whether a species is extinct or at risk of becoming extinct. The assignments are based on the current population of a species and any known threats to its population. Threats can include things like habitat loss, **poaching**, problems with breeding, and more. Conservation status categories are meant to help people decide which species need protection the most.

How did different organizations categorize lake sturgeon in 2024? Look

Big, Odd, and Old

Sturgeon have swum in the rivers, oceans, and lakes of our planet's northern hemisphere since the time of dinosaurs. There are twenty-seven species of sturgeon worldwide. Some live in salt water and others live in fresh water. In Wisconsin, only two species of freshwater sturgeon cruise the bottoms of our waterways: lake sturgeon and shovelnose sturgeon. Shovelnose sturgeon are found only in at the table to find out. (Each organization uses its own list of conservation status categories. You can visit their websites to find out which categories each organization uses and how they are defined.)

Conservation Categories			
Organization or list	**What area does it cover?**	**How did it categorize lake sturgeon in 2024?**	**Where can I learn more?**
International Union for the Conservation of Nature (IUCN) Red List of Threatened Species	Worldwide	Endangered	iucnredlist.org
NatureServe	US and Canada	In Wisconsin: Vulnerable In other US states and Canadian provinces: Vulnerable, imperiled, critically imperiled, and presumed **extirpated**	explorer.natureserve.org
US Endangered Species List	US	Not listed (not threatened or endangered)	ecos.fws.gov/ecp
Wisconsin Natural Heritage Inventory (NHI) Working List	Wisconsin	State status: Special concern	dnr.wisconsin.gov/topic/NHI/WList

the Mississippi River **watershed**, which includes the Mississippi, Wisconsin, Chippewa, Red Cedar, and St. Croix Rivers. Historically, lake sturgeon lived throughout Wisconsin and were found in all waters that drain into Lake Michigan, Lake Superior, and the Mississippi River. Now, lake sturgeon are less common everywhere. In some places, they're long gone. But in Wisconsin's Lake Winnebago system, they're still healthy and thriving. And **spawning**.

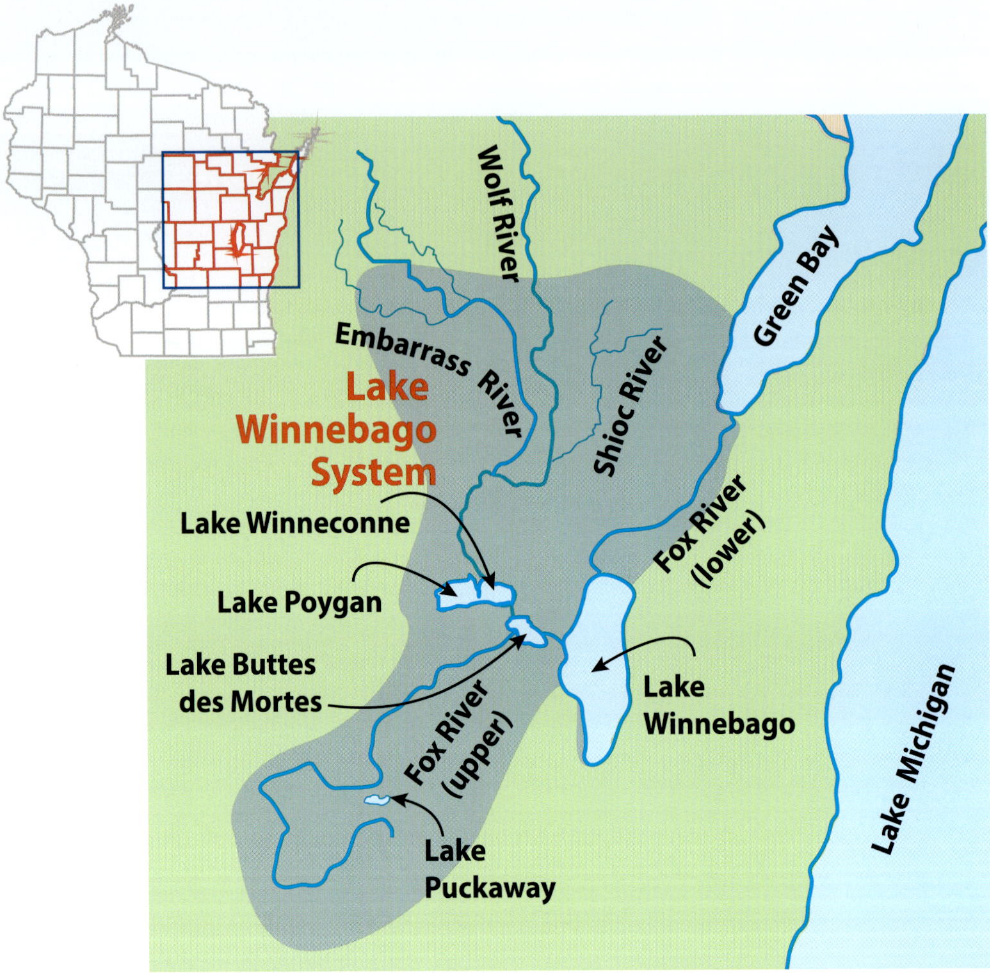

Back on the Wolf River, the just-released lake sturgeon zooms upriver with powerful flicks of her tail. She weaves between, around, and under the other sturgeon that crowd the waters. Her **dorsal fin** sometimes pokes up above the water like a shark's.

Also like sharks, sturgeon don't have bony skeletons. Their skeletons are made of softer **cartilage**. Their cartilage skeletons do not include a backbone, like other fishes'. Instead, a cartilage-wrapped rod of nerves runs down the length of a sturgeon's body from head to tail fin (also called a caudal fin). This is called a **notochord**. All fish at one time had notochords, but most fish species evolved to have the bony backbones we see today. Not the sturgeon.

Sturgeon don't have scales like most other species of fish. Instead, rough, grayish-brown skin covers a sturgeon's body. It feels a little like fine sandpaper. Five rows of long, bony plates called **scutes** run along their backs and sides for protection. A young sturgeon's scutes have points that are sharp enough to cut your hand, but they wear down and stretch out with age until they just look like rounded bumps on their skin. As the biggest fish in the river, mature lake sturgeon don't need the scutes to protect themselves from predators. Mature sturgeon can grow as big as six feet long and weigh two hundred pounds. Not only are sturgeon the biggest fish in Wisconsin, but they also live the longest. They can live for more than a hundred years in the wild. The oldest sturgeon ever caught in Wisconsin was eighty-two years old. Just think—there could be sturgeon swimming in our waters that were around before your great-GREAT-grandparents were even born!

A mature sturgeon with worn, barely visible scutes

Sturgeon Anatomy

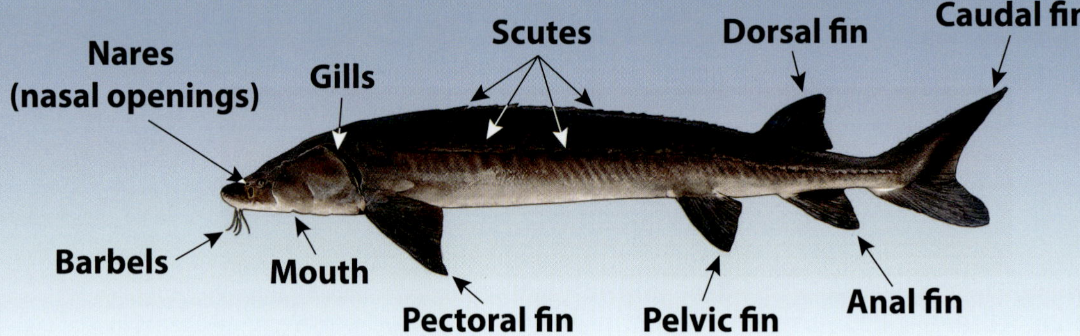

Stadig, the sturgeon biologist, gushes over them. "These fish are fantastic," she says. "This is one of the living dinosaurs. They are a swimming contradiction. They get so big, but they don't have any teeth. They don't feed like a predator. They basically swim around eating bugs."

But our sixty-two-inch female sturgeon, just released back into the Wolf River, doesn't care about eating right now. She's on a mission to release her eggs onto the rocky banks of the river. This is her first time returning to the waters where she was born.

Because sturgeon live so long, it takes them a long time to mature, or become ready to spawn. Most other fish are ready to reproduce when they are three to four years old, and then they release eggs every year. Female sturgeon are twenty-five years old before they are ready to reproduce, and even then, they spawn only every five years or so. Males are ready sooner, but it's still fifteen years before they make their first trip to spawn. Mature males can spawn every other year.

Several male sturgeon take notice of our female as she approaches. They thrash and flick their tails over the rocks. They're releasing **milt**, or sperm, into the water. In response, she spills hundreds of thousands of eggs that stick to the rocks. The eggs that are fertilized by the milt will soon begin developing into young sturgeon.

Sturgeon eggs are vulnerable at this stage. Many are eaten by other fish and crayfish. Those that survive will hatch in five to eight days. In two weeks, the young sturgeon will be an inch long and ready to start searching for their own food. However, survival remains tough. Although their mother released thousands of eggs, only about twenty of them will survive to adulthood.

Young sturgeon, also called juveniles, look so different from adult sturgeon that at first people thought they were two different species. Juvenile sturgeon's scutes are spiky. Their noses are pointy and turned-up. Their bodies are thin and sharp looking, with wide side fins that stick out like ears. It's only after years of survival that they thicken and round out to resemble their parents.

The sturgeon who manage to survive spend their days coasting along the bottoms of rivers and lakes. They grow whiskers, called **barbels**, that they drag along the bottom of the waterway. Sturgeon use these highly sensitive organs to find food. As Stadig mentioned, they have no teeth, so they digest everything whole. Their barbels bump along the sand, mud, and stones. When they come across small snails, crayfish, leeches, or insect **larvae**, the

In the spring, sturgeon move upriver to spawn. Hundreds of fish thrash through the water.

Juvenile sturgeon look very different from their parents, but both juvenile and mature sturgeons' mouths are on the underside of their heads so they can suck up their food from the bottom of the lake or river.

sturgeon's tubelike mouth extends and sucks up the food like a vacuum cleaner. They suck up the sand and gravel, too, which they spit out through their gill openings.

Back in the river, the female sturgeon's job is done once her eggs are released. She coasts downstream, past the other sturgeon swimming upstream to deposit and fertilize their own eggs, past the crowds of people pointing and admiring along the shore. She won't return to this river again for five years.

This spring ritual—for the sturgeon and the people watching along the river—lasts only for a few days. If the weather turns too warm or too cold, the spawning season will wrap up early. Everyone watching these ancient fish today knows that they're lucky to catch this event.

As Stadig says, “I can’t think of many other species where you get people who don’t even fish for them coming out to celebrate this species. A massive amount of people just come to watch them spawn. They’re not catching them—they can’t do anything with them.” They just want to see them.

Suddenly, another sturgeon across the river rears up out of the water, as if he wants to make sure everyone’s trip to the Wolf River was worth it. Splash! He falls back on the river’s surface. No one knows why sturgeon do this. Maybe it’s to knock off parasites. Maybe it’s to signal something to the other fish. Whatever the reason, the crowd gasps and cheers, glad to have witnessed these amazing ancient fish.

CHAPTER 2

FROM MILLIONS TO THE BRINK OF EXTINCTION

Objects like this mussel shell decoy and bone harpoon tell us that Indigenous people have been harvesting sturgeon for centuries.

Long before crowds drove to Shiocton to marvel at the splashing sturgeon, people celebrated the fish in a different way.

For thousands of years, the Native peoples of the Great Lakes region depended on and honored the lake sturgeon. The Menominee Nation has an especially important spiritual connection to the fish. Sturgeon were also a crucial source of food. Often, it was sturgeon that helped the **Indigenous** people of the land that became Wisconsin survive the long winters.

In winter the Menominee fisher would make a hole in the ice, then lay a blanket of evergreen boughs next to the hole, where they would lay on their stomach to watch for sturgeon. The fisher would also build a small, teepee-like structure over the hole to block out the light and avoid scaring the fish. This would also improve visibility for the fisher. The fisher might dangle a decoy shaped like a sturgeon down the hole to catch a sturgeon's attention. Then, the fisher used a spear or harpoon to catch any sturgeon that swam past.

Fishing for sturgeon through the ice is one thing, but imagine the joy and relief when the ice finally melted and masses of sturgeon surged up the rivers to spawn in the spring! Native people celebrated—and stocked up.

Native peoples harvested sturgeon **sustainably** for thousands of years, allowing the fish's population to thrive. Before the 1800s, it's estimated that eleven million sturgeon swam in Lake Michigan alone. That's twice as many sturgeon in one body of water as people who lived in Wisconsin in 2020! Wisconsin waters were once so thick with sturgeon that in the springtime it was hard to cross the rivers. A machinist and farmer from Berlin, Wisconsin, described the fish on the Fox River in a magazine article:

> During these few days of their run you could see these fish six or seven feet long run up out of the water their full length and fall in again with a huge splash, sometimes three or four at a time, during this time it was not safe to sit on the railing of the bridge unless you wanted to get soaked. One big fish following another would lunge out of the water and falling back would splash the water clear across the bridge. . . . Horses were frightened and runaways were common.

Despite all this, by the late 1800s sturgeon were in big trouble in much of Wisconsin. By 1914 the leader of the US **Commission** of Fish and Fisheries wrote, "The story of the sturgeons is one of the most distressing in the whole history of the American fisheries. It requires no prophet to foretell that in a comparatively few years the sturgeon will be practically extinct."

From Despised to Prized

What happened? How did we go from rivers full of giant, leaping fish that scared the horses to possible extinction?

It began in the early 1800s when European settlers first started pushing into the land now known as Wisconsin. These settlers had a radically different view of the fish than the Native peoples did. While some of them probably found themselves, like the Native peoples, depending on sturgeon for a meal, most of these newcomers despised the fish. European settlers who fished on the Great Lakes especially hated sturgeon because their sharp scutes often tore holes in their fishing nets. If they came across sturgeon, they'd often drag them out of the water and leave them on the shore to die.

Some European settlers believed sturgeon to be a useless fish that ate the eggs of other fish, so they clubbed them to death. And that's not all. They burned them for steamship fuel. They fed them to their pigs. They buried them in farm fields to help their crops grow. They did everything they could to try to get rid of them.

That is, they tried to get rid of them until the late 1800s. By then, trains were able to transport blocks of ice to keep foods cool, and the fancy Russian food **caviar** became popular. Guess where the best caviar came from? Sturgeon. Suddenly, the black eggs from Wisconsin's lake sturgeon were considered "black gold." Now the settlers couldn't get enough sturgeon. Cities sprang up along the Great Lakes to process the fish. In 1872, just one sturgeon processing plant on Lake Erie churned through more than thirteen thousand fish and almost seven hundred thousand pounds of caviar. They shipped products from the fish—caviar, smoked sturgeon meat, and isinglass (gelatin made from the fish's bladder)—all over the world.

By the late 1800s, butchering floats like these were common across the Great Lakes. Butchers sliced open sturgeon for their caviar, then tossed the rest of the fish away as garbage.

Lake Winnebago sturgeon fishers and their harvest in 1913

James Milner, the Great Lakes deputy commissioner for the US Commission of Fish and Fisheries in the 1870s, championed this new use of sturgeon. Instead of lake sturgeon being plowed into fields or left to rot, he noted that "out of a shameful waste of a large supply of food they have established a large and profitable industry."

Unfortunately, too many people embraced the industry. Without any rules and regulations to abide by, people overharvested the sturgeon. As a result, sturgeon production went from 7.84 million pounds in 1879 to 1.77 million pounds just twenty years later, in 1899. The decline was not due to there being less interest in caviar. As fisheries commissioner Hugh M. Smith said in 1914, it was due to "the most reckless, senseless fishing imaginable, with the result that in a comparatively few years the best and most productive waters were depleted." Simply put, sturgeon were being harvested and destroyed at a rate that far exceeded the birth of new sturgeon.

And that's not all. Something else happened during the 1800s in Wisconsin that harmed the lake sturgeon population. Even though it wasn't done to the fish directly, it impacted them immensely. People began building dams.

Darn Dams

As new cities and towns and farms and industries popped up all over Wisconsin, people also built dams on rivers throughout the state. Rivers were dammed to provide power to cut lumber and grind grain and light up towns, to help with flood control, to water crops, and to create lakes where there were none.

The problem with dams, though, was that sturgeon need miles and miles of open river to spawn. The dams stopped the sturgeon from swimming to their spawning areas. The sturgeon were literally gated off from hatching new fish.

Even when some sturgeon managed to release eggs in a dammed river, the conditions were often poor. The water was cloudy with sawdust or pollution, not clear as the sturgeon needed it to be. Also,

This dam on the Potomac River in Maryland in the 1890s had a **fishway**, or ramp. The ramp gave fish a way to move upstream past the dam. Most dams built in the 1800s in Wisconsin did not include fishways.

spawning below the dam simply didn't give the young sturgeon enough space and time to grow safely before drifting out to the wider, deeper waters of the lakes. As senior DNR fisheries biologist Aaron Schiller puts it, "Sturgeon will pass decent spawning habitat lower in the river because their larvae have to drift for a certain period of time, and the young fish typically live in the rivers for the first summer, and then migrate out in October. So, if they spawn lower in the river, those small fish have a shorter area to drift and a smaller area to survive."

The Danger of Dams

Wisconsin law defines a dam as "any artificial barrier in or across a watercourse which has the primary purpose of impounding or diverting water." The first dam in Wisconsin was built on the Fox River in 1809 to power a sawmill, which was used to cut lumber. Many, many dams followed, all over the state. As of 2024, Wisconsin had almost four thousand dams on its waterways, yet many needed repairs or weren't necessary anymore.

Sturgeon are not the only wildlife harmed by dams. Dams change the water's quality, temperature, and oxygen levels. All those changes affect the living things that use the water, from birds to microscopic organisms to humans.

The Little Rapids dam on the Fox River in 1900

One Safe Place

All this damming was happening at the same time that people were killing sturgeon for being a "nuisance" and then overfishing them for caviar. But it was a while before anyone noticed a problem.

Do you remember how slowly sturgeon grow? They don't spawn until they're between fifteen and twenty-five years old. Even then, a female sturgeon lays eggs only every five years. That slow growth meant that it took a while for humans to notice changes in the sturgeon population. When people finally did notice and took action to help, it would still take the fish a long time to reverse course—a hundred years or more. Most sturgeon populations in the Great Lakes region have not yet recovered from the destructive 1800s.

However, there was one tiny pocket in Wisconsin where lake sturgeon avoided the worst of the problems other sturgeon encountered. The Lake Winnebago system, which includes Lake Winnebago, the Upper Fox, Wolf, and Embarrass Rivers, and the connected smaller lakes of Butte des Morts, Winneconne, and Poygan, has everything sturgeon need. The sturgeon that live in these waterways don't have to migrate out to other waters. And these rivers went mostly undammed because lumberjacks needed to float the logs they cut down the currents to Oshkosh. This meant the rocky, clear rivers in the Lake Winnebago system were available for spawning, while the wide, shallow lakes were perfect for mature fish to cruise and suck up larvae and crayfish. It was a perfect, closed system.

Ironically, dams helped to further protect the Lake Winnebago sturgeon population. In 1856, fourteen dams and seventeen locks were built on the Lower Fox River, which connects Lake Winnebago to Green Bay and Lake Michigan. These dams effectively sealed the sturgeon into the Lake Winnebago system, which helped the fish avoid the rampant pollution, loss of habitat, and overfishing that occurred on the bigger lake.

So, Wisconsin had its own somewhat self-contained lake sturgeon population. Unfortunately, not everyone in Wisconsin had the same idea about what would be best for the Lake Winnebago sturgeon. Many of the disagreements to come would revolve around one big question.

CHAPTER 3

WHOSE FISH ARE THEY, ANYWAY?

By the late 1800s, it became obvious to all who used Wisconsin's waterways that the numbers and size of all species of fish—not just sturgeon—were quickly dwindling. The construction of dams, cutting down of forests, overfishing, and industrial pollution had all taken a toll. But how could people change course and ensure that Wisconsin's fish would not only survive, but thrive? That question sparked a debate that continues to this day.

To start protecting the state's fish, in 1874 the Wisconsin state government got involved. Government officials organized the first statewide Fisheries Commission. Members of the commission set about visiting the rivers and lakes to see just how bad the situation was and to come up with ideas for how to make things better.

Their first idea was simply to **stock** the lakes and rivers with fish. It's a little like when something runs out at the grocery store and the

The Peshtigo Company's dam and sawmill, pictured here, blocked sturgeon and other fish from swimming up the Peshtigo River to spawn in the 1890s.

store orders more to replenish the shelves. The Fisheries Commission started building fish **hatcheries** so that they could raise fish and replenish Wisconsin waters.

But as commission members visited the lakes and rivers, they saw that the situation was worse than they thought. Stocking was one way they could help, but it was not going to solve all the problems the fish were facing. Wisconsin's fish needed more protection.

There Ought to Be a Law

In 1879, state superintendent of fisheries H. W. Welsher asked the state **legislature** to create laws to protect the sturgeon. "The evidence of the decline of the fisheries is daily growing more manifest, and only by our labors . . . can we hope to arrest its decay," Welsher said. He saw the dams and the spearing of fish under the dams as catastrophic.

At that time, there were some laws regarding fishing in Wisconsin, but they had been created at the county level, not for the entire state. You could be fishing on a lake or river and, depending on where you were on that body of water, the rules for the number, size, and species of fish you could take would be different. It was impossible for anyone to keep track of the laws, much less follow or enforce them.

The next task of the Fisheries Commission was to compile all the different county fish laws into one document. But when they brought this document to the state legislature to make it into law, arguments erupted. People wanted the ability to set the rules for their area of the state, but sometimes their rules contradicted their neighbors' rules. No one wanted to give in or compromise.

Finally, after much wrangling, the legislature established some basic fish laws for Wisconsin. Almost immediately, people in different regions of the state protested and pressed their state **representatives** to make exceptions to the laws. It soon became a battle between what Barney Devine, the state's chief conservation warden, called "well organized and noisy group[s]" that wanted control over their own waterways and how they were fished, and conservation groups, like the Fisheries Commission and the Izaak Walton League, that wanted statewide laws.

A Heated Dispute

The "noisy groups" believed that their rights to their own local land and waters were being trampled. They wanted to be able to fish and enjoy the recreational traditions of their families. The state conservationists concentrated on the bigger picture. They knew that the state's fish were not headed in the right direction and needed safeguarding. They worried that without broad intervention, the fish in Wisconsin would disappear just as other natural resources had, such as the old-growth forests that had been clear-cut during the 1800s logging boom, or the passenger pigeons, which had gone from millions to extinct in just a few years due to overhunting and loss of habitat.

The two sides were separated by a fundamental difference in how they viewed the state's natural resources. Were fish, wildlife, and nature the property of local owners or of the state as a whole? Who did the fish belong to, really?

It was hard for the conservationists to defend their perspective since they couldn't offer evidence to back up their concerns. The state didn't know any specifics about the sturgeon population, only that it seemed as though fewer were around. It was a little like parents telling their kids they needed to do something "because I told you so." And anyway, people were reluctant to believe things were bad across the state when *their* fishing was just fine. Frustrated, the conservationists called these noisy groups "fish hogs" because they believed these groups were considering only themselves and their own situations. There was a lot of bickering back and forth.

Many Wisconsin families build or purchase small shelters called ice shanties that they haul onto frozen lakes in winter. Shanties are placed over a small hole in the ice used for fishing. The walls protect ice fishers and spearers from the wind. Many people take great pride in their ice shanties, like this boy in 1898 on the bay of Sturgeon Bay.

Local groups in the Lake Winnebago area were especially outspoken regarding sturgeon. After all, *their* sturgeon weren't scarce like in the rest of the state. They eventually pushed through exceptions to many of the state fishing laws that had passed earlier in the 1880s that restricted certain fishing practices. In 1887, spearing was made legal on Lake Winnebago. In 1892, gill nets were allowed. In 1899, ice **shanties** were permitted on the lake.

Fishing for Sturgeon: Setlines, Snag Lines, Gill Nets, and More

By the late 1800s, people had devised ways of catching large quantities of fish with the least amount of effort. Some used gill nets, setlines, or snag hooks that allowed them to do other work while the net or hooks caught the fish.

Gill nets are long swathes of nets, some as long as two hundred yards, that hang vertically in the water. Fish swim into them. Their heads go through the holes in the net, but their bodies can't fit. They get caught by their gills. Sturgeon scutes would sometimes tear this type of net, ruining it.

Fishers in Maine reel in their gill nets to dry and store them.

Fishers also used setlines. They ran a rope across the river with lines, hooks, and weights dangling from it. These lines would hook fish that swam by. Setlines were very effective during spawning season, when thousands of fish moved up and down the rivers. In addition, people often snagged for fish by casting out a heavy hook and jerking it to snag a passing fish.

These fishing methods concerned early conservationists. The problem with all of these methods is that they catch anything that swims into them, no matter the size and no matter the species. It's also difficult to return unwanted fish caught by these devices back to the water alive.

People dangle sturgeon snagging lines from a bridge over the Wolf River in Waupaca around 1870.

Versions of all these methods are still used today in the oceans and smaller waterways in some parts of the world, and they're still a cause for concern for the same reasons they were over a hundred years ago. Today in Wisconsin, gill nets are banned for the regular fisher, but researchers and commercial fisheries can use them.

Drastic Measures

In 1915, the Fisheries Commission was folded into the larger statewide Conservation Commission. The Conservation Commission was tasked with overseeing *all* of the state's natural resources, and it set about making fishing regulations for the good of the whole state, not just the "noisy groups," as conservation warden Barney Devine had called them. The new Conservation Commission took a drastic step for sturgeon in particular: a complete statewide ban on all sturgeon harvesting.

As you can imagine, the people who had harvested sturgeon for generations were upset, especially those in the Lake Winnebago area where sturgeon populations were not as diminished as in other places. Now, that tradition was being taken away. An article in a 1916 edition of the *Oshkosh Daily Northwestern* bemoaned the new law, saying "a big fat sturgeon was seen rubbing his back against one of the pilings just below the Huron street bridge Monday afternoon. It was about nine feet long and looked as though he'd make good feed. But the law says no."

Many people broke the law and fished, netted, or speared sturgeon anyway. Newspapers at the time were dotted with articles about arrests and fines for poaching sturgeon. "Illegal fishing on the Wolf is so well known that it is a common thing for a fish hog to be dubbed a 'Winneconne sportsman,'" joked William Mauthe, chairman of the Conservation Commission, referring to the fishers in a small town on the Wolf River.

Some broke the law because they believed it was unfair. Others believed they had no other choice. They needed the fish for food or income. Ed Gorchals, who was from a small town near Lake Winnebago, was arrested for poaching. He remembered, "They took my boat and motor and my license for the year, put me in jail for three days." When he was asked why he did it, he said simply, "We needed the money." Robert "Jake" Abraham, who grew up on the west shore of Lake Winnebago, agreed. "That's how we made extra money," he said. "There was no money in farming."

There were some, however, who schemed to get rich with the new law. Lee Ogden was arrested for packing poached sturgeon in trunks and sending them by train to Chicago, where they were bought and sold for a profit. When Prohibition—the nationwide ban on the sale of alcohol—started in 1920, some people made a lot of money making and selling liquor illegally. Some of the same people also sold illegally harvested sturgeon and caviar.

Then the Great Depression hit in the 1930s. Many people lost their jobs. People went hungry. Many thought it was time to change the sturgeon ban so people could feed their families. People thought, too, that surely the sturgeon were doing better now, after fifteen years of no harvesting. It was time to relax the law.

The Conservation Commission thought that was short-sighted. In 1931 Paul Kelleter, the conservation director, wrote to the governor that loosening the restrictions on sturgeon fishing would "doubtless hasten the complete destruction of sturgeon in Winnebago waters.

This fisher netted a six-foot sturgeon on LeClair pond in 1934.

Sturgeon has not increased in numbers in these waters sufficiently to warrant an open season at this time and the removal of protection as contemplated by the proposed legislation before you is not in the best interests of conservation."

Nevertheless, state legislators from communities around Lake Winnebago drafted a bill to open a hook-and-line fishing season in fall and a two-month spearing season in winter. The bill passed into law in June 1931. This was the start of the winter spearing season on Lake Winnebago that people know today, although there have been changes since 1931.

More Disagreement, More Debate

About the same time, there was a push to move the power to set fishing and hunting regulations from the state legislature to the Conservation Commission. The constant jockeying from different parts of the state to update and tweak laws was gumming up Wisconsin's government. Besides, most legislators didn't really know much about fish, anyway. Shouldn't rules about fishing be left to the fish experts?

Frank N. Graass, from the Izaak Walton League conservation group, explained what deciding on fish regulations looked like when it was left to the Wisconsin legislature:

> During every session of the legislature, the legislators introduced bills for themselves or at the request of some individuals or group of hunters or fishermen irrespective of the merits of the bill. Every session 1,200 to 1,400 bills were introduced and an average of 200 bills per session related to fish and game. . . . These bills included changing seasons, the methods of hunting and fishing, length of

> fish, [and] trapping laws, relating to specific counties or county. Consequently, the rules and regulations were so confusing that it almost took a civil engineer or surveyor to tell you where you could hunt and fish and maybe a Philadelphia lawyer to tell you what the law was.

In 1933, the Wisconsin legislature finally turned over control of fishing and hunting rules to the Conservation Commission. Now it was up to just one group of people who were trained in conservation to make the decisions about fishing and hunting laws for Wisconsin. But they also accepted input from a public advisory board called the Conservation Congress. The congress was made up of a representative from each Wisconsin county. However, the congress could only give recommendations, not make decisions.

In 1938, a debate bubbled up over ice fishing in general. The Conservation Commission worried that people were taking too many fish under the cover of their ice shanties. They worried that fish numbers weren't recovering as they had hoped. Wisconsin resort owners joined in their concerns over ice fishing. After all, fewer fish meant fewer tourists in the summer.

When the Conservation Congress met and discussed the ice fishing issue, winter sturgeon spearing soon got lumped into the debate over ice fishing in general. Spearers fished in shanties, too, and who knew what really happened inside those little shelters?

The old argument resurfaced: Who did the fish belong to, really? Local folks who lived on or near the water? Or the state as a whole?

Both sides spoke up. A conservation warden in Shawano County said, "From an enforcement angle, it is impossible to regulate. They [the public] will cooperate a lot more if there is no ice fishing on sturgeon."

Howard Leppla, a Conservation Congress representative from Calumet County, argued, "We have the only sturgeon in the state of Wisconsin in quantities large enough to spear. We want our sturgeon spearing through the ice."

The Conservation Commission members listened. They took notes. Then they did their own thing. They announced that hook-and-line fishing for sturgeon would end on January 15 and spearing would be reduced to one month, January 15 to February 15 only.

In 1947, wardens confiscate nine sturgeon that had been illegally snagged from the Wolf River.

The spearing community around Lake Winnebago was furious. Locals organized petitions and protests. The commission explained that it had heard them, it just didn't agree with them:

> The fact is well known that fish and game are the property of the state and the conservation commission acts as custodian of such fish and game. It is not reasonable to assume that the fish and game, whose habitat is in the vicinity of certain towns and villages, are any more the property of the individuals of that vicinity than any other resident of this state. Likewise

> the commission feels that the fish in Lake Winnebago are the property of the entire state rather than subject to other ownership of the residents of that particular locality.

The protesters did not agree. "It is our lake; we fish on it, and we know the conditions that exist out there," one fisher countered at a public hearing in Fond du Lac. Still, the Conservation Commission did not consider this a good way to make resource management decisions.

Then Leppla explained to the crowd that he had recently asked Edward Schneberger, the fisheries biologist on the Conservation Commission, how the group had come to its decision. Leppla said:

> I asked Dr. Schneberger for the following information: the estimated sturgeon population in Lake Winnebago, the number of sturgeon removed yearly from Lake Winnebago for the last ten years, and the estimated increase or decrease of sturgeon in those waters for the past ten years.
>
> Dr. Schneberger said it was impossible to answer these questions, but after all, if we could arrive at some means whereby we could get some sort of a fish **census** it would be a great help.

With that speech, a way for both sides to move forward together became clear. Maybe if the commission could provide the spearers and fishers with data and explain the reasoning behind its decisions, those fishers would be more likely to support the commission's rules and laws. Instead of saying, "Do it because I said so," the commission would give fishers reasons to do what it asked.

The problem was, the commission didn't have that data. At least, not yet.

The Lake Winnebago Sturgeon Spearing Tradition

If you aren't part of the tradition, it probably seems like a weird thing to do: go out onto a frozen lake in the middle of winter, cut a hole in the ice, pull a little shed-like structure called a shanty over the hole, then spend hours staring into it, hoping a sturgeon swims by so you can throw a giant sharpened fork at it. Many people do this for years without catching a single sturgeon. If you do make a successful catch, then you have to wait in line at the outdoor registration station with your now-frozen, bloody fish in a sled to have it weighed and measured.

But for generations of people living near the Lake Winnebago system, sturgeon spearing season is the highlight of the year. It's a time for families and friends to get together, celebrate their triumphs, admire the fish, and tell stories about the ones that got away.

Two successful Lake Winnebago spearers in the 1950s.

The tradition of spearing for sturgeon in the winter came from the Native peoples who have lived in the area for thousands of years. But it has continued in a way that probably no one imagined. Now, for about two weeks in February every year, hundreds of trucks, cars, utility task vehicles, and snowmobiles dot the snowy ice of the lake. Spearers use the vehicles to tow out their ice shanties, which can be as simple as garden sheds or as comfortable as campers with heat and Wi-Fi.

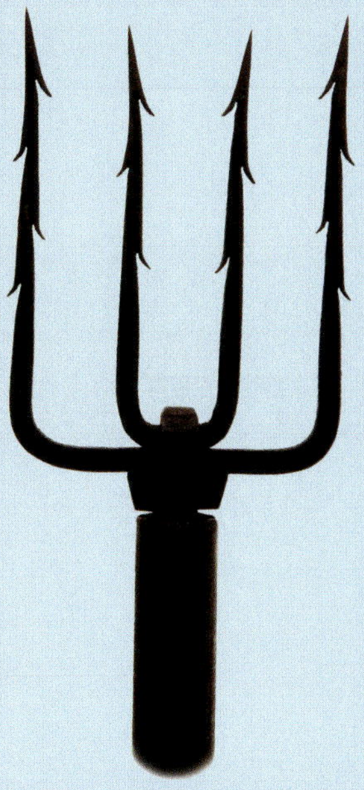

Spearers use chainsaws and electric augers to cut holes in the ice. They lower PVC pipes down to the bottom of the lake to spot the shadow and the size of a sturgeon more easily as it swims over the pipe. Decoys of all kinds are lowered down to lure the fish. Spearers use beautifully carved wooden decoys, pie pans, coffee mugs, and even children's toys. Often a spearing group has its own superstitions and recommendations for how to attract fish. However, anything put into the lake must be removed when spearing season is over.

The shanties are kept dark so spearers can more easily see into the water. But there's plenty of snacks, music, and drinks to help make the wait more fun. Spears are at the ready for the moment a sturgeon happens to swim under the hole. Spears are six to eight feet long, with barbed tips that won't slip out if the sturgeon jumps or twists to get away.

In recent years, the use of underwater cameras to more easily spot the fish has been a growing issue for the spearing community. According to a survey conducted in 2013, most people saw cameras as giving spearers an unfair advantage and wanted them banned from shanties. But a more recent survey in 2021 suggests this may be changing. In general, younger spearers tend to support the use of technology, while older generations tend to oppose it.

Whether cameras will be banned is still to be determined. Either way, the next generation of spearers are most definitely looking forward to sitting out on the ice in February.

Sturgeon decoys and spears come in many shapes and sizes, reflecting each individual spearer's personal tastes and preferences.

CHAPTER 4

STURGEON BY THE NUMBERS

What in the world is a "fish census"?

That's what Howard Leppla asked the Conservation Commission for in 1939. In other words, Leppla wanted an accounting of the number, age, and location of the lake sturgeon in Wisconsin. It sounded like a great idea and a logical way to begin making sensible fishing rules and regulations.

But how exactly do you count sturgeon? The Conservation Commission had been wrestling with this question for decades already. During the late 1930s, the commission again fell back on the idea of stocking sturgeon. Raising sturgeon from eggs in hatcheries would allow biologists to observe the fish growth and habits, and they could easily count the number of new sturgeon being released into the lakes and rivers.

It seemed easy, but it proved to be anything but. When the Conservation Commission began to investigate how to raise sturgeon in hatcheries, it could not find anyone who had ever been successful at it. The fish always died. Some died before they even hatched, usually when a fungus invaded the eggs. The fish that hatched died by the time they grew to about one inch long at twelve to fourteen days old. In those cases, the young fish refused to eat.

"I am rather firmly convinced that the reason our fish died when they did was due to lack of proper food," wrote a professor in Toronto about his attempt to raise sturgeon. "No hatchery or any other method will restore or even maintain a stock of fish as sensitive to environment and as slow to grow and reproduce as the sturgeon of our waters."

This grim news convinced the commission not to go the hatchery route.

So then what?

The Fishers Lend a Hand (and a Fin)

If stocking more fish wasn't possible, then the commission needed to make sure that the existing sturgeon in the Winnebago system were not overharvested, as they had been in other areas of the state. But to answer Leppla's questions about how many sturgeon there were and how many were harvested, the commission needed to answer some even more basic questions. How long do sturgeon live? How old do they need to be to start spawning? Only by gathering this information could the commission start to estimate the numbers of lake sturgeon in the Winnebago system—and from there, make informed decisions about sturgeon spearing and other regulations.

In 1941, Edward Schneberger, the man who couldn't answer Leppla's questions (and Wisconsin's first state fisheries biologist), had an idea. Fishers were required to purchase a tag, or license, to catch a sturgeon each year. What if the commission used that registration tag to ask for information about the sturgeon being caught?

The 1969–1970 postcard for Menominee River sturgeon fishers. The data were used to help make future decisions for the species.

MICHIGAN - WISCONSIN
MENOMINEE RIVER SURVEY

Please complete this card when you are through fishing today by recording (1) the number of hours which you fished, (2) the kinds and numbers of fish which you caught and kept, and (3) the size of sturgeon. You can help us, yourself and other fishermen by filling out the card and mailing it soon.

(1) NUMBER OF HOURS FISHED ☐☐

(2) YOUR CATCH

KIND	NUMBER

(3) SIZE OF STURGEON

LENGTH	POUNDS	TAG NO.

☐ I did not catch any fish.

MO. ☐☐ DAY ☐☐ YEAR ☐☐ BOAT ☐ SHORE ☐ SECTION: A ☐ B ☐

MICHIGAN - WISCONSIN
MENOMINEE RIVER SURVEY

MO. ☐☐ DAY ☐☐ YEAR ☐☐

Mr. ☐
Mrs. ☐
Miss ☐

(Fisherman's Name)

(Number and Street)

(City) (State) (Zip)

BOAT ☐ SHORE ☐ SECTION: A ☐ B ☐

Schneberger made up some postcards to go with the tag. He asked the people buying the tags to fill out the postcard and send it back to the commission. The postcard asked for the length and weight of any sturgeon caught.

But that's not all Schneberger asked for. He also asked people to send in their sturgeon's head.

Why? He wanted to learn more about the age of the fish that were being harvested. One way to determine a fish's age is to look at a part of its ear called an **otolith**. Biologists can estimate the age of a fish by counting the growth rings in the otolith, a little like the way you can count the rings in a tree's trunk to estimate its age.

A cross section of the tiny otolith in a sturgeon's ear. Counting the number of rings in the otolith is the most accurate way to tell the fish's age.

Surprisingly (or not), many people didn't want to send in their sturgeon's head. But not because it was gross or messy or hard to mail. They wanted to keep the head to make soup out of it. Making sturgeon head soup was and still is a tradition for many sturgeon-spearing families.

That first year, only fifty spearers sent in their sturgeon heads. But many more returned the postcards, enough to start gathering some data on the fish. Biologists learned that sturgeon take over twenty years to grow old enough to lay eggs, which meant protecting them was even more vital. There would be no quick recovery if something happened to reduce the Lake Winnebago sturgeon population.

In 1953, the Conservation Commission's fisheries biologists modified Schneberger's plan. Instead of asking for the sturgeon's head, they asked spearers to send in the front, or pectoral, fin. This is another way that scientists can age a fish. Just like with the otolith, if you cut a fish's fin, you'll find countable age rings.

A cross section of a sturgeon's pectoral fin. Counting fin rings can be an effective way to tell the age of a young sturgeon. However, the older the fish, the less accurate this method becomes.

In return for the fins, the biologists offered to report back to the spearers the age of their fish. Suddenly, it was a win-win situation for all. Spearers got to keep their sturgeon heads. They also got information back that was interesting to them. The Conservation Commission got what it needed, too: more samples and more data. The program was a huge success. In 1953, spearers turned in 966 fins. The effort became one of the best and biggest sturgeon studies in history. It was also a huge lesson for everyone involved. Seemingly opposing sides really could work together! Continuing to do so might just be the way forward for safeguarding the future of sturgeon in Wisconsin.

The 5 Percent Limit

The fin collection and aging of the fish was just one part of a comprehensive study of the life of lake sturgeon in the Lake Winnebago system in the 1950s. The Conservation Commission also started tracking the migration of the fish from the lakes to the rivers to spawn. In the spring, biologists would wade into the rivers, net the fish, measure them, and attach a metal tag to their top fin before releasing them. These tags had unique numbers on them, giving the biologists a way to track a fish's movements over time.

At left, scientists attach a metal tag to a sturgeon's dorsal fin. Though these tags (also pictured above) have not been used in years, some sturgeon swimming in Wisconsin waters today still have them—evidence of how long sturgeon live.

With this new data coming in, the biologists started to construct a picture of the sturgeon population in the Lake Winnebago system.

The commission modified its fishing regulations with the new information. It shortened the spearing season to three weeks and allowed each spearer only one fish per year. It started requiring spearers to register their catches. This way officials could check for metal fin tags and gather more information. Over time, the commission could estimate the number of sturgeon swimming through the system that were big enough to catch. Those numbers helped the commission develop sustainable numbers for harvesting each year.

Gordon Priegel, a fisheries research biologist hired in 1959, came up with a harvesting standard: the 5 percent limit. It was a simple concept. Based on the number of sturgeon estimated in the system each year, the next year's harvest was limited to 5 percent of that total number. While the exact number of sturgeon might change from year to year, the impact on the population would always be the same and predictable. Today, because biologists have more data on sturgeon lifespans, they are able to use complex models to determine the harvest cap each year, but it was Priegel who first set the idea into motion. "I figured it was conservative, and I figured it would work," he said.

With all these new developments and more reliable numbers, people began to relax about the sturgeon population in the Lake Winnebago system. The Conservation Commission slowly phased out the spring tagging. The registration stations, formerly run by the commission's biologists, were turned over to tavern owners. As you might expect, the numbers and the data began to get foggier.

By 1967, when the Conservation Commission became the Wisconsin Department of Natural Resources, government officials thought it was no longer necessary to have a fisheries biologist dedicated to sturgeon. After all, the sturgeon population of Lake Winnebago was under control. Sturgeon were safe.

Or were they?

CHAPTER 5

STURGEON FOR TOMORROW

By the late 1960s, there was no more spring sturgeon tagging. Private businesses such as taverns had taken over the reporting of the spearing information. The role of fisheries biologist dedicated to sturgeon had been eliminated.

As data collection about the state's sturgeon slowed, the number of sturgeon speared each year also declined. Was it because fewer people were spearing? Were they less successful? Were harvests being reported inaccurately? Or maybe the changing harvest caps affected the data? Whatever it was, the decline in number of fish speared was dramatic. Numbers dropped from an all-time high of 2,828 in 1953 to eight—just eight!—in 1969. While the reason for the drop wasn't clear, there was one potential factor that was quite literally unclear: the water.

Bill Casper stands outside his Green Bay Packers–inspired ice shanty in 1998.

After years of runoff from roads, farms, and paper mills and even sewage spilling into the water, the clarity of the Lake Winnebago system's lakes and rivers had declined dramatically. The increase of pollution caused algal blooms, reduced the amount of light that could get through the water, and killed the aquatic plants. It's impossible to spear sturgeon if you can't see into the water, no matter how many fish are (or are not) in the lake.

By 1977, all these things worried Bill Casper, a lifelong Lake Winnebago fisher. He watched the numbers of speared sturgeon decrease every year. Was something catastrophic happening to the lake or the fish? How would the sturgeon survive? In Casper's opinion, it was crucial that Wisconsinites figure out how to raise sturgeon outside of their lake environment, in a hatchery.

Members of the newly formed DNR dismissed Casper's idea of trying to raise sturgeon in a hatchery. "We had the life history and population, [so we thought] we didn't need any more," explained Dan Folz, the new fish manager at Oshkosh at the time. Members of the DNR also recalled the investigation from the 1930s that had ended when the Toronto professor told them raising sturgeon was hopeless. But sometimes, as Folz said later, "you think you know almost everything, but then you find out how much you don't know."

Rebuffed by the DNR, Casper started talking to spearers at the local taverns about raising sturgeon to repopulate the lakes. They didn't need convincing. The tavern owners were worried about the sturgeon numbers, too. Soon, it became an organized effort. Casper and the tavern owners held fundraisers and named their group. "What you're trying to do is raise sturgeon . . . for tomorrow," said Casper's Uncle Henry one night. With that, the Sturgeon for Tomorrow group was born.

Before long, Casper approached the DNR again. DNR employees agreed to a meeting at the Taycheedah Town Hall, where they were joined by 150 spearers.

The DNR still balked at the idea of trying to hatch sturgeon. It had never been successful. Plus, there was a concern about how it would look to outsiders. If they raised sturgeon to stock the lakes, wouldn't that be admitting there was a shortage? Members of the DNR worried that if the federal government heard about the declining population, it might declare the fish threatened or endangered. Then there would be no fishing or spearing for anyone.

Eventually, the spearers posed a new idea: What if they didn't raise the sturgeon for stocking the lakes, but raised them just to learn how to do it, in case there was a natural disaster of sorts someday that wiped out the lakes' fish?

"We entered [the meeting] with an open mind," Jim Addis, the DNR fisheries chief, reflected years later. "If Sturgeon for Tomorrow could produce a workable program, we'd be glad to assist."

The "Recipe"

Sturgeon for Tomorrow already had a hatchery location in mind. The retired Reverend Walter Lang ran a private trout hatchery called Wilderness Springs in New London, Wisconsin. He volunteered his tanks and ponds.

Now they just needed to figure out how to do it. It wasn't that no one had ever tried raising lake sturgeon. Some had. Every single time, all of the sturgeon died because of a fungus that grew on the eggs or because the hatched sturgeon wouldn't eat. How could this time be different?

Bill Casper had another idea. He had read an article in *National Geographic* magazine about Dr. William Ballard from Dartmouth College in New Hampshire. Ballard had been studying how Russians were raising other sturgeon species. Casper gave him a call.

Ballard was incredulous. In those pre-internet days, he hadn't realized that there was a thriving sturgeon population in the United States. "My God, man, where are you calling from?" he asked. "You have sturgeon in Wisconsin? I've been going to Russia and Romania to study sturgeon."

Ballard was in. He would help in any way he could. The local community would, too. Sturgeon for Tomorrow held fundraisers, and people stepped forward with money for the experiment, eager to help. One business lent a thousand dollars and asked the group to pay it back

Dr. Ballard was stunned to learn that, yes, there are sturgeon in Wisconsin!

when and if they could. It took years, but Sturgeon for Tomorrow did eventually pay back the loan.

The vice president of Sturgeon for Tomorrow used his own money to fly Dr. Ballard and his wife, Elizabeth, to Wisconsin from New Hampshire in February 1978. Together Dr. Ballard, the Wisconsin DNR, and the Sturgeon for Tomorrow members gathered at a hotel in Fond du Lac.

Ballard sat in the middle, with Sturgeon for Tomorrow members and DNR staff around him. "I know the recipe for this," Ballard said. He explained that Russian scientists had worked with Ossetra, Beluga, and Siberian sturgeon for fifteen years before figuring out how to mix the eggs with red clay so that the eggs didn't stick together. The clay was the key to keeping fungus from growing on the sturgeon eggs.

The next morning, Ballard came out of the hotel with a piece of paper for the Sturgeon for Tomorrow members—a "recipe" for raising sturgeon in a hatchery. "Guard this with your life," he said. Sturgeon for Tomorrow turned it over to the DNR, and the first attempt to raise lake sturgeon in forty years was underway.

The First Attempt

First, the team had to get some fertilized sturgeon eggs. Dan Folz, the Oshkosh fish manager, headed out with two colleagues to one of the sturgeon spawning spots. The three men nicknamed themselves Egg Man, Pan Man, and Clay Man. Folz, the Egg Man, had the job of getting the eggs from the female sturgeon. Pan Man Mike Primising was in charge of collecting the eggs in a pan and fertilizing them with milt from a male sturgeon. Clay Man Russ Daley would add that extra ingredient—clay—that everyone hoped would keep the eggs from

Fred Binkowski, a UW–Milwaukee scientist, gathers milt from a male sturgeon in 1980 or 1981.

sticking and developing the fungus that had killed the sturgeon eggs in previous attempts.

They netted a female early in the night, but she wouldn't be ready to release her eggs until a male was nearby. Ballard had explained that when the Russian scientists took the sturgeon eggs, they simply killed the female and harvested the eggs from her. The DNR men didn't want to harm a fish intentionally. So they kept the female on a canvas tarp on shore while they waited for a male sturgeon to swim by. They dumped buckets of water over her to keep her wet.

It was hours before the men finally netted two male sturgeon. When they did, the Pan Man and Clay Man leapt into action. They pressed on the bellies of the fish to collect the eggs and milt in the same pan. Then they stirred and mixed. Now they could return the fish to the river. The males took off immediately. Because the female sturgeon had been out of the water for so long, Egg Man Dan had to swish her in the water for about forty-five minutes before she finally splashed and swam away. The team was relieved to have successfully collected fertilized eggs without harming any sturgeon.

By that time, it was morning. Getting the fertilized eggs had taken all night! But they weren't done yet. The three DNR men raced to Reverend Lang's hatchery, where they moved the eggs into troughs and trays.

Reverend Lang led the men back out to his kitchen. "Now we can relax for a minute," he said. "You guys have been up all night, so it's time for a cup of coffee." The tired men sat back and celebrated their success.

Unfortunately, the success was short-lived. Reverend Lang kept daily notes on the eggs. Even though the scientists had used Ballard's clay recipe, the eggs started growing fungus. Eventually, all of the eggs died.

The men were crushed, but they vowed to try again.

The Second Attempt

The next year, in 1979, the DNR had even more resources for a second egg trial. Instead of Reverend Lang's hatchery, the group would use the Wild Rose State Fish Hatchery, a trout-raising hatchery established in 1908.

When spring rolled around, it was time to head back to the river and catch some spawning sturgeon. This time Don Czeskleba, the

An employee inspects incubator jars at the Wild Rose State Fish Hatchery sometime during the 1960s or 1970s.

Wild Rose DNR hatchery supervisor, joined the Egg Man, Pan Man, and Clay Man on the shore. Also present was Fred Binkowski, senior scientist at the University of Wisconsin–Milwaukee. The DNR would share some of the eggs with Binkowski's lab.

Eight-day-old sturgeon eggs in a McDonald jar, which is designed to rotate the eggs. Inside the jar are eyed eggs (eggs that have developed enough that their eyes are visible) and sac fry (newly hatched fish that still use their egg sacs as a food source).

Again, the researchers caught a female sturgeon first. But this time, instead of keeping her on a piece of canvas until they could catch a male, Czeskleba surprised Folz by taking out a small knife. "What are we doing?" Folz asked. Well, Czeskleba knew that many sturgeon survived spearing attempts, so he figured that a small cut to remove the eggs from the fish wouldn't harm her, especially if he stitched her right back up. And that's what he did. "I don't think she was out of the water . . . fifteen minutes," said Folz. It was a lot faster than the hours-long process from the previous year!

They brought some of the fertilized eggs back to the Wild Rose hatchery and sent the others to Binkowski's UW–Milwaukee lab. The men must have been on the edge of their seats, wondering what would happen—none more so than Dr. Ballard, the Dartmouth scientist who had shared the Russian "recipe." He had been invited back to Wisconsin for the event. He even slept on the floor to keep watch over the eggs.

Everyone was thrilled when the eggs hatched! But the challenges weren't over. Now they had to keep them alive for longer than two weeks.

Binkowski, the scientist from UW–Milwaukee, had been studying research from the 1930s on what young sturgeon ate. He remembered how the professor from Toronto was convinced that their fish had died because they didn't get the right food. So Binkowski experimented. He tried commercial fish food. Nope. He tried frozen brine shrimp. Nothing. The fish wouldn't eat.

Then he tried live brine shrimp. The baby sturgeon slurped them up.

"It must have something to do with the nutritional value or the smell, or something that's still there to say to them, 'This isn't frozen and it hasn't been sitting in a freezer for six months, this is freshly killed and I can eat this, I want this,'" Binkowski explained.

Slowly, the young sturgeon grew. Eighty percent of the eggs hatched and grew into full-sized lake sturgeon. The experiment was a rousing success. A first in the world for lake sturgeon!

Czeskleba wrote up the details of the experiment and published them in a journal for the scientific community. Lots of people must have read the article because, suddenly, people from all over the world wanted to know more. Fishery staff from other states visited and toured the Wisconsin hatcheries. They took sturgeon eggs back to their hatcheries, hoping to restart lake sturgeon populations in their states.

Even More Research

This success launched a new era of sturgeon research. In 1990, when Dan Folz retired, Ron Bruch became the new Oshkosh area fish manager. One of the first things he did was restart the springtime netting, measuring, and tagging of sturgeon on the rivers.

In the years since the 1950s, when the sturgeon were first tagged, new tracking technology had been developed. Instead of the bulky metal tags biologists clamped to sturgeons' fins, now a tiny glass PIT tag full of coiled copper wiring could be inserted under the sturgeon's skin at the back of its head. These tags transformed how fish were tracked. Since the tags were smaller and stayed under the sturgeon's skin, they were less likely to be lost or damaged. The fish still had to be captured for the PIT tag to be scanned, but the PIT tags were more accurate than the old metal ones. Instead of having to manually read numbers on a thrashing fish, a person could swipe a small handheld scanner over the fish, activating the tag and sending the fish's unique identification number to the scanner. PIT tags are still used widely today. In fact, if you attend any spring sturgeon spawning event on a Wisconsin river, you'll see biologists and volunteers scanning for the tags and inserting them.

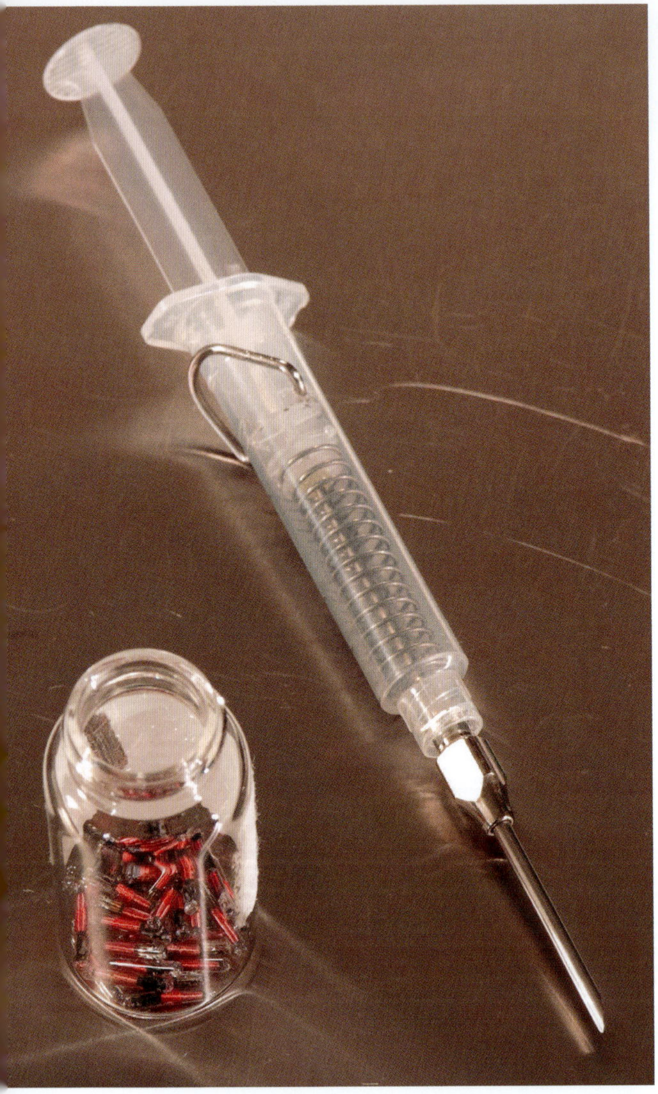

PIT tags (at left, in a glass jar) are less than half an inch long and inserted into animals with a syringe (at right).

Bruch also knew there was more to learn about the differences between male and female sturgeon. For years, the number of large female sturgeon being speared compared to males was higher than the DNR would like. It was important to protect the large, mature females because they would create the offspring that would keep the sturgeon population healthy. However, no one knew why more females were being caught than males. Raising the size limit to protect the big females didn't seem to fix the problem.

Bruch and his team studied the fish coming into the spearing stations and determined that female and male sturgeons mature at different rates. It takes females longer to mature, so they are bigger than males when they are first ready to spawn. When the DNR raised the size limit for sturgeon harvesting, they were accidentally targeting the females. Once the DNR understood that, they lowered the size limit, and fewer mature females were speared.

Bruch also figured out how to determine the age of a sturgeon more accurately. Since 1953, people had been using the front fins and counting the rings to age a sturgeon. But by measuring the radiation in the carbon in sturgeons' otoliths, Bruch found the fin data wasn't always accurate. It turned out that Ed Schneberger, the fisheries biologist who had asked for spearers' sturgeon heads years earlier, had had the right idea. The otolith is the more accurate way to age a sturgeon.

With this new information, Bruch updated the DNR's database of sturgeon, recalculating the age of fourteen thousand fish. This in turn solved a mystery.

For years, biologists had noticed that in their database of captured sturgeon, there were fewer sturgeon in a certain age range. They had data for older and younger sturgeon, but there was a time span during the 1920s when significantly fewer sturgeon appeared to have been born. Why? This had bothered the biologists for years.

With the correct aging, however, Bruch could now safely assume that many more fish had been born in the 1920s than experts had assumed. They would have been just reaching adult size levels in the 1930s, a time when poaching with setlines and snag lines was rampant. It wasn't that fewer fish had been born during that time, but that many more had been poached.

It was a sobering realization. “It’s a testament to how far back you have to look in the past to find out what could have affected the current population,” Bruch said. “Sturgeon aren’t really managed on a yearly basis, they’re managed over decades and centuries.” In other words, the consequences of decisions made and actions taken today probably won’t be apparent until decades into the future. This long game also makes it harder to reverse course if a mistake is made in managing the sturgeon population. It’s a tricky business.

CHAPTER 6

NAME AND NAMA'O

The idea that sturgeon management needs to happen over decades and even centuries was not new to the Indigenous peoples in Wisconsin. Native peoples had lived along rivers and lakes in the area for thousands of years before European settlers arrived. Throughout that time, Native peoples fished and harvested sturgeon sustainably. In fact, sturgeon were essential to many Native nations' diets and cultures.

The Ojibwe people originated in the Great Lakes area, migrated east, and then returned about 1,500 years ago to the northern part of what is now Wisconsin and Minnesota. Their migration path follows the waterway in which lake sturgeon flourished historically, and the Ojibwe culture honors the fish. The sturgeon (name in the Ojibwe language) is the Chief of their Fish Clans. The Ojibwe call the Namekagon River Namekaagong-ziibi, which means "river at the place

A carved sturgeon decoy made in 1996 by John V. Snow, a member of the Lac du Flambeau Ojibwe. For generations, Ojibwe fishers have lowered decoys like these through the ice in winter to lure sturgeon close enough to spear.

abundant with sturgeon." Ojibwe people still harvest sturgeon today using their traditional methods. They're able to fish on lands both on and off their reservations due to the nations' treaty rights, which were negotiated in 1837 and 1842.

The Menominee people's connection with sturgeon in the land now known as Wisconsin goes back ten thousand years. As Dave Grignon, the Menominee Nation's historic preservation officer, says, "We have a name for ourselves, 'Kiash Matchitiwuk,' which means 'the Ancient People' because we have been here for so long. We have no migration story. Our creation story took place only sixty miles east of the present reservation."

During those thousands of years, the Menominee lands stretched over ten million acres. And sturgeon, called Nama'o in the Menominee language, were at the center of the people's cultural, spiritual, and nutritional lives. Along with wild rice and maple syrup, sturgeon were a staple in their diet. The Menominee people speared them through the ice in winter and by canoe when the waters were open.

Historically, the Menominee dried or smoked the fish to make it last longer. They used the whole fish for all sorts of things. "People used it for dancing, some of the parts. But a lot of it was for medicine. Certain medicines were extracted from the sturgeon," Grignon explains.

There was also a traditional celebration each spring, when the ice melted and the sturgeon stormed up the Wolf River to Keshena Falls, where they would spawn. Grignon describes it. "One of the first things that the Menominees did after Creation was fish sturgeon on the Menominee River," he says, "and thanked the Creator for that fish. After the first catch of sturgeon, there was a great ceremony and feast. A feast and singing and dancing. Ceremonial prayers all went along with that." Like the Ojibwe, the Menominee have a clan structure. It is the responsibility of the Menominee sturgeon clan to organize and lead the ceremonial sturgeon tradition.

Traditions Destroyed

As European settlers began moving into the Menominee Nation's lands, Menominee culture, celebrations, and sturgeon were gradually taken from them. In 1848, the new state government of Wisconsin pressured the Menominee to give up their lands and move to a reservation in Minnesota near the Crow Wing River. Menominee leaders traveled to the proposed site to see if it could work. But they returned, declaring, "the poorest region in Wisconsin was better than the Crow Wing," and besides, a conflict between the Dakota and Ojibwe Nations made it unsafe for the Menominee to live in Minnesota. They took their objections to the federal government and the US president in Washington, DC. Eventually, the Menominee's protests, petitions, and delays convinced the state to let them stay on a sliver of land

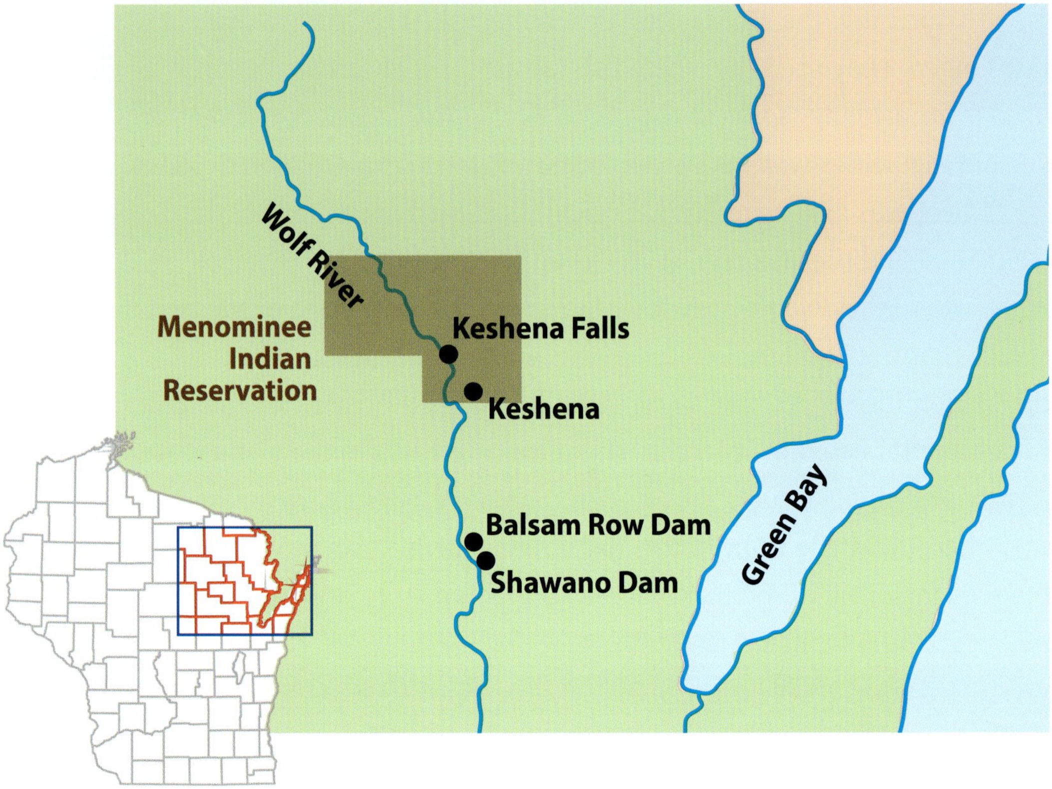

in Wisconsin that included Keshena Falls and the Wolf River, so the Menominee people could still honor some of their cultural traditions.

In 1854, the people who once hunted and fished freely across ten million acres of land were allotted 236,000 acres. But, thanks to the Menominee Nation's leaders, the Wolf River and Keshena Falls were part of the new reservation. The people's sturgeon traditions continued. For a while.

Can you guess what happened to the Wolf River? Flip back to chapter 2 for a hint.

That's right, the Wolf River between Lake Winnebago and Keshena Falls on the reservation was dammed. In 1892, several men received

Bear Trap Falls near Keshena in 1929. Because of the dams on the Wolf River, sturgeon were no longer found at places like these within the Menominee Reservation.

permission from the state to build a dam at Shawano to power their mill and manufacturing businesses. In 1926, a second dam called Balsam Row was built even closer to the Menominee Reservation. Both were constructed without a fishway or passage for the sturgeon. Both were constructed without any consultation with the Menominee Nation. Even though a part of the river was within the Menominee reservation, their beloved sturgeon could no longer reach them.

The dams made a hard situation even worse. For a while, some Menominee people traveled off of the reservation to catch sturgeon in other rivers during the state's legal spearing season. Unlike the Ojibwe, the Menominee's treaty rights did not include off-reservation harvesting beyond what was legal for the general population. So when the Conservation Commission banned all sturgeon harvesting in Wisconsin in 1915, it meant there was nowhere the Menominee could connect with sturgeon and their cultural traditions. They were stuck.

When the state sturgeon regulations began to allow some fishing and spearing in the 1930s, Menominee leaders approached the state to ask about the fish. Menominee elders attended a 1930 Indian Affairs meeting that had been organized by a local community group in Oshkosh. Representatives of the Conservation Commission were also at the meeting. Pete Pamonicutt, a Tribal court judge, began:

> We used to have sturgeon on the reservation where we came from, and I know that the dam has prohibited the sturgeon from coming up on our reservation. Now can it be possible that sturgeon be planted in some lake, even if there are only a dozen, so we can put them in our big water? My ancestors used to come to Lake Winnebago early in spring when snow was still around on the ground, and get a supply of sturgeon and take them to our present reservation. We also had a good supply before the dams were established. It is always surprising to us that they did not put in a fishway in Shawano.

The Conservation Commission must have agreed, because there was a transfer of fish in 1935 and again in 1944. But it wasn't enough. Fifty years of the state government preventing sturgeon from migrating up the river to spawn had taken its toll on the population.

The last sturgeon harvested during that era on the Menominee Reservation was speared at Keshena Falls in 1959. Another thirty years went by before things began to change for the better.

The Sturgeon Return

"In 1992 a group of Menominees, including myself, got together and said, we need to get these sturgeon back," Grignon recalls. "We're missing out on the ceremony. It's part of our culture. They are spiritually connected to us. Let's get these back."

But how?

"We were going to go down to Shawano and harvest a sturgeon," Grignon says.

To be clear, that would be illegal. Spearing is not allowed on the Wolf River at any time.

Grignon continues: "Our Tribal chairman said, 'No, you guys better not do that. You can get arrested. Let me talk to the state and see what I can do to get some sturgeon for your ceremonial feast,' which he did. And the DNR brought fifteen sturgeon to us for the feast and ceremony. And ever since then, the Tribe is getting these ceremonial fish to do our sturgeon feast and celebration, along with the ceremony that goes along with it."

A Menominee elder who still remembered the sturgeon ceremony from decades before performed it at Keshena Falls that first year. The nation has had a spring sturgeon celebration ever since. People perform ceremonial dances, feast, and celebrate. "It's just a happy time like it was thousands of years ago when the Menominees would see the sturgeon in the spring," Grignon says.

Menominee Tribal Historic Preservation Officer David Grignon sprinkles tobacco and offers blessings after the sturgeon are delivered by the DNR while other members of the Menominee Nation watch.

Don Reiter began a new position as the fish and wildlife manager for the Menominee Nation during this time. He was involved in determining the plan to return sturgeon to the reservation. Reiter explains, "We had federal, state, and Tribal biologists all come together for initial talks. We finalized the plan in 1995."

There were a few key parts to the initial plan. In addition to the ceremonial fish the DNR delivered, the Menominee elders asked for a "look and see." They wanted Menominee people to be able to come and

Menominee Tribal School students transfer ceremonial sturgeon from a DNR tanker truck to a holding pond on the Wolf River.

see sturgeon spawn at their original spot: Keshena Falls. So the DNR agreed to bring sturgeon up above the dams by tanker truck and release them. In 2012, for the first time in more than one hundred years, sturgeon spawned in their ancestral site.

Another part of the plan was "put, grow, take." The DNR and the US Fish and Wildlife Service began stocking **fingerling** and **yearling** sturgeon in a lake on the Menominee Reservation. In 2005, that stocked lake's population became strong enough that the Menominee opened a harvest season.

Keshena Falls, or Nama'o Uskiwamut, is an ancestral spawning grounds for sturgeon and a place of spiritual and historical significance for the Menominee Nation.

The Menominee also track and study the sturgeon. That's where Reiter comes in. He tags the fish with PIT tags as well as acoustic tags. The acoustic tags allow him to track the fish's movements by listening to signals sent from the tags to a receiver. The Menominee are learning more about sturgeon behaviors and habitats from these methods.

However, because the Menominee went so long without the sturgeon, a lot of Menominee people today don't know the traditions. A lot of teaching is happening. About how to fish. About the ceremony. About the fish themselves.

And, the fact is, sturgeon are arriving at Keshena Falls via truck, not naturally. As Douglas Cox, the director of land management for the Menominee Nation, says, "We want to see sturgeon making it from Lake Winnebago to Keshena Falls, and we want to be able to celebrate that every spring."

More Work to Be Done

One reason the current plan is frustrating to the Menominee is that less than half of the sturgeon delivered above the dams stay in the Upper Wolf River. With acoustic **telemetry**, biologists can tell that more than 50 percent of the sturgeon find their way back downstream and get stuck there again.

The Menominee would like to see permanent **fish passages** added to the Shawano and Balsam Row dams so the sturgeon can make their way to Keshena Falls on their own. But the Menominee don't own the dams, which aren't on their land. Nevertheless, in 2006, the Menominee won a lawsuit against the operators of the closest dam. The suit's settlement says that a fish passage must be added. However, it's the Menominee Nation's responsibility to pay for it. The solution has been tangled up in conflicting views for years, and still no fish passages have been added.

One of the holdups comes from the DNR's concerns about opening the Wolf River. The agency, which oversees dams in Wisconsin, has concerns that invasive species and disease in the Lower Wolf River might spread to the fish in the Upper Wolf system. The DNR would prefer to keep delivering and stocking the fish at the reservation. If a fish passage is built, the agency wants it to include a facility for examining and sorting the fish. That costs a lot more money to build and maintain, and so far, fish passages have not been included in the DNR's water management plan for the Upper Wolf River.

Still, the Menominee Nation presses on, applying for grants and funding to afford the more complex project that the DNR requires. The nation wants to fix both the "ecological disconnect" that the dams create for all fish as well as the cultural disconnect. "We want our history, we want our culture back," Cox says.

He explains that it's not about harvesting sturgeon on the river at all. Current state and Menominee Tribal fish regulations prohibit that. As Cox says, "I don't see us changing that. We're not interested in harvesting sturgeon on the river. Our culture is what's being denied. The fish should be moving up and down that river without man's interference. Menominee culture isn't going to be satisfied until those fish make it to Keshena Falls on their own."

The Menominee Creation Story and Menominee Clans

Dave Grignon, the Menominee historic preservation officer, tells the story of the Menominee people's beginnings:

> Thousands of years ago, at the mouth of the Menominee River, the creation story took place of the Menominee Tribe. And it was passed down from generation to generation. That's how we know the creation story.
>
> It was said that the Great Bear came up out of the mouth of the Menominee River and started walking up the river. And the Creator saw him and turned him into human form. That was the first Menominee. As the Great Bear was walking up the river, he saw Eagle flying above. He said, "Brother Eagle, come down and walk with me. Be my brother." Eagle descended and walked with Bear, and the Creator saw that and transformed him into human form, also—the second Menominee. As the two walked up the Menominee River, they encountered a moose, wolf, and crane. Those were transformed into human form, and those are the five major clans of our people.

The Menominee have thirty-five clans that are grouped into the five principal clans of Bear, Eagle, Moose, Wolf, and Crane. Each Tribal citizen is born into a clan, and each clan has its own responsibilities. Grignon explains:

> The Bear clan are the orators and leaders, where the chiefs come from. The Eagle clan is more for making laws, freedom, and justice. Moose clan are the protectors of the wild rice [and oversee its harvest]. The Wolf clan has the responsibility of sharing the hunting and fishing and gathering activities for the Tribe. And the Crane clan are artists and architects and the ones who provided our structures for our people. Sturgeon clan is the keeper of the wild rice. They're under the Bear clan. The Sturgeon clan are the keepers of telling when to harvest sturgeon. They are the ones that direct the ceremonies for the feast and celebration.

CHAPTER 7

HATCHING STURGEON TODAY

Remember those fingerling and yearling sturgeon that are stocked in the lakes on the Menominee Reservation? Where do they come from? Biologists gather the eggs and milt from sturgeon on the Lower Wolf River in April or May when the fish spawn. Then they take the eggs and milt to hatcheries: the US Fish and Wildlife Service's federal hatchery in Genoa, Wisconsin, and the Wild Rose State Hatchery in Wild Rose.

That's right. Ever since that first successful experiment hatching sturgeon back in 1979, sturgeon continue to hatch and grow in Wild Rose.

While the first fish hatcheries in Wisconsin were outdoor ponds that were open to the weather and predators, now they are highly technical indoor/outdoor facilities, driven by science, research, and funding. Hatcheries in Wisconsin are run by the state, by private owners, by Tribal governments, and by the USFWS. These hatcheries produce all kinds of fish native to Wisconsin waters.

Raising sturgeon in tanks at the Genoa National Fish Hatchery in Genoa, Wisconsin

What Happens at Wild Rose?

At Wild Rose, a tiny town along the Pine River, the hatchery is now a state-of-the-art facility that sprawls across two sides of the highway. Wild Rose is a cold- and cool-water hatchery, raising not only sturgeon but also muskellunge (also called muskie), walleye, salmon, and trout.

"Typically, we raise several different **strains** of sturgeon here," explains Sam Stauflien, Wild Rose fish culture biologist. "We take eggs from the Fox and Wolf Rivers, as well as Lake Winnebago and the Milwaukee River. Each location has a little different water chemistry, and the fish will imprint on that and return to those waters to lay future eggs." Stauflien and his coworkers take careful steps to ensure the fish survive.

To start, anyone who steps into the building must dip their shoes in a shallow tray of disinfectant and squirt their hands with sanitizer. Biosecurity is a serious business in a hatchery.

The raising of sturgeon begins in the egg disinfectant room. "We call this our drive-up window," Stauflien says. "I ask for eggs, and the technician behind the counter will joke, 'Order up!' as they slide a beaker of fish eggs through the window in the wall."

The eggs have already been through several steps before they arrive at the hatchery. First, they are fertilized. The eggs from one female are fertilized with milt from five males. At least eight females are fertilized this way. "Eight is a good number for **genetic diversity**," Stauflien says.

Then the eggs are swirled with a mix of water and diatomaceous earth, a clay-like gardening product. When sturgeon eggs are fertilized, they become sticky. In the wild, this helps them to stick to rocks along the shore. In the hatchery, though, it causes them to ball up in the tank, not get enough oxygen, and die. (This is exactly what happened during early efforts to raise sturgeon.) The clay of the diatomaceous earth bonds with the sticky substance and allows the eggs to separate enough to get the oxygen they need.

The clay coating gives the eggs a light brown color. It also enables them to be washed with iodine, a liquid used to kill germs. They're cleaned twice—once at the harvest site, and then again in the egg disinfectant room. This is to prevent the spread of the deadly fish disease viral hemorrhagic septicemia (VHS).

Sturgeon don't get VHS, but they can spread it to other fish. If eggs or water carrying the virus were introduced to the hatchery, it would be disastrous. One year, a wild muskie that the hatchery spawned tested positive for VHS. "We had to **euthanize** about thirty thousand eggs from that female," Stauflien says. He explains that if any fish tested positive for VHS at the hatchery, "by law, we'd have to kill everything off in this building. We would have to euthanize 400,000 to 500,000 fish. Taking these precautions keeps us from having to do that."

Once the eggs are cleaned, they leave the egg disinfectant room and are placed in rows of small vertical tanks on the main floor to start growing. The lights are kept dim in the hatchery since the fish spook when light hits them. The cavernous room is a quiet hum of motors and water recirculating.

What is VHS?

Bulging eyes. Bleeding. Bloated abdomens. Death. These are all symptoms of viral hemorrhagic septicemia (VHS), a deadly fish disease that first arrived in Wisconsin in the early 2000s. Infected fish spread the virus in the water through their urine and other fluids. Other fish swim through the virus particles or eat infected fish and catch the disease, and the cycle starts again. While lake sturgeon aren't harmed by it, VHS is deadly to other important Wisconsin fish such as muskellunge, largemouth bass, perch, and trout. Because of that, scientists are working hard to contain the disease. So far, VHS has been limited to the Lake Winnebago system, Green Bay, and the Great Lakes. It has not been found in the rest of Wisconsin's waterways.

How can you help prevent new infections of VHS in fish? Be a responsible boater and fisher. Just like the hatchery biologists, be careful about mixing water from one source into another. The invisible-to-our-eyes virus could be lurking in that puddle in your boat or your minnow bucket. So, drain your boat and fishing containers when you leave a body of water, and don't transport live fish (including minnows) or fish eggs from lake to lake or river to river. Following these guidelines can help stop the spread.

Containers of live brine shrimp that fish hatchery staff raise to feed the sturgeon hatchlings

After about a week, the sturgeon start hatching. They nibble on the live brine shrimp that the hatchery staff raises for them. Remember Fred Binkowski, the scientist from UW–Milwaukee who figured out back in the 1970s that sturgeon like fresh, not dried, stuff? His findings hold true today.

As the sturgeon grow, they are moved into tanks the size of an aboveground pool. Because sturgeon are bottom feeders, the tanks don't need to be filled to the top as they are for the other fish in the hatchery.

Their food changes here, too. At first, hatchery staff chop up bloodworms for them to eat. But once the sturgeon are four inches long, they are ready to start snacking on **krill**, which is the same food that whales in the ocean eat.

Throughout this process, the hatchery staff watch over the growing fish, trying to make sure that as many sturgeon as possible survive. One thing they watch out for are fish bullies. Sturgeon grow at different rates, just like people do. Sometimes the bigger fish will bite or take food away from the smaller sturgeon. This is when the hatchery staff steps in.

"We have to do what's called 'grading' them," Stauflien explains, holding out something that looks like a tray with long, narrow openings across the bottom. "We'll put fish in there, and obviously, the big ones will stay behind, and the small ones will go through [the openings]. We'll keep all the big ones together so they don't beat on each other. There are times where those fish will get to a big enough size difference where they'll actually try to eat other ones." He gestures to a stack of differently sized graders. "We'll have to grade them a half a dozen times throughout their life cycle or else we lose a lot more."

Sturgeon hang out in the hatchery for longer than any other fish at the Wild Rose facility. Some are released as fingerlings in the fall after they hatch. But most spend the winter in the hatchery. In February, they are shipped off as yearlings to places like the Menominee Reservation, where they're released through holes cut into the ice.

Scientists learning how to successfully rear sturgeon in a hatchery changed the game for sturgeon populations all over

Wisconsin. Rivers and lakes where sturgeon were scarce or even extirpated (like on the Menominee Reservation) now have sturgeon swimming in them, thanks to the hatchery.

From 1983 to 2001, fingerling sturgeon from Wild Rose were added to the St. Louis River, which flows into Lake Superior in the far northwestern corner of Wisconsin. Wild Rose sturgeon also swim in the Namekagon River.

The sturgeon reared at Wild Rose travel as far as Kentucky, Missouri, and Tennessee, too. Lake sturgeon have been long gone from those states, but fisheries biologists there are hoping to bring them back. Wisconsin's healthy sturgeon are rebooting the species' population in rivers around the United States.

The Sturgeon Life Cycle

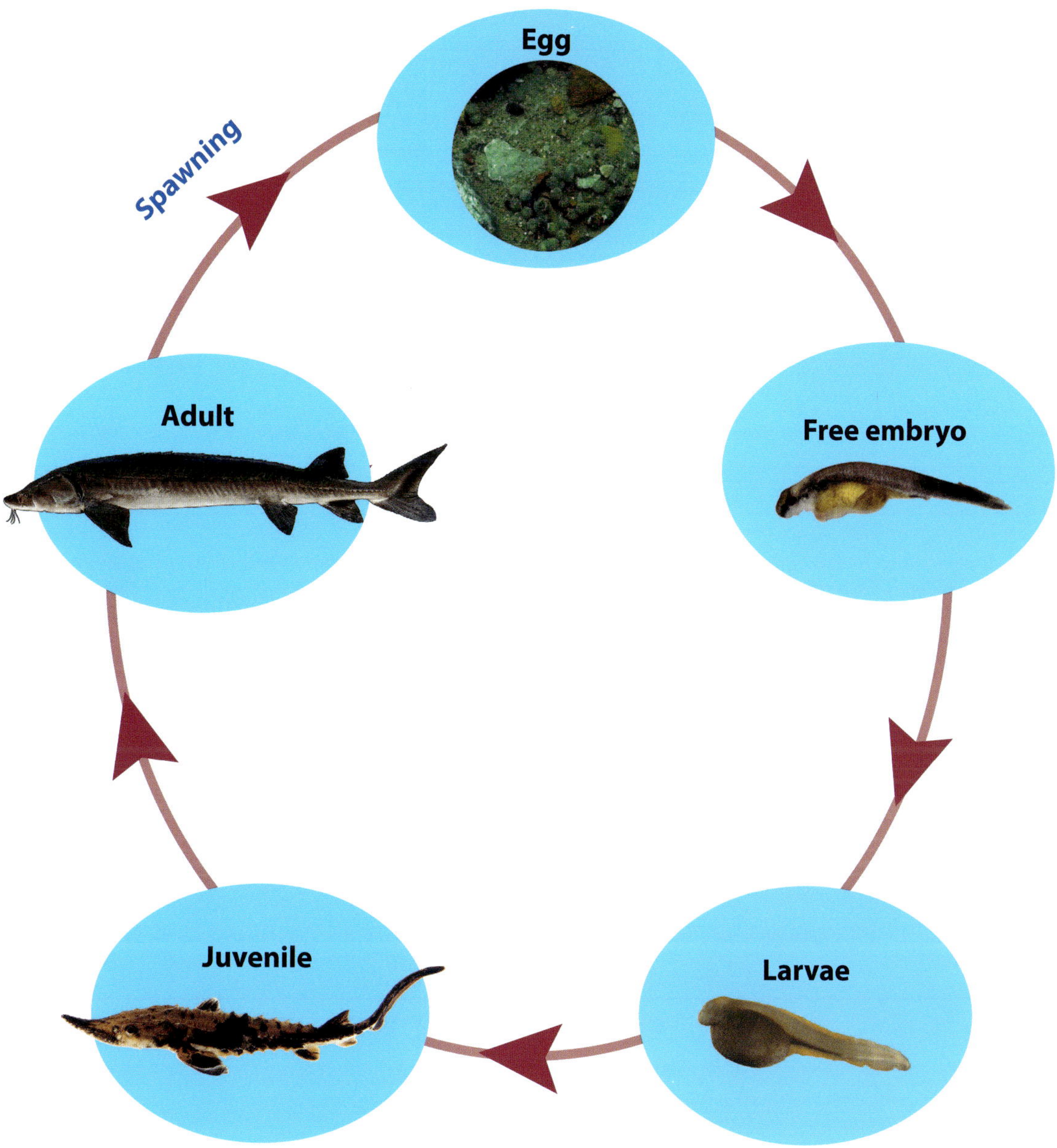

Mobile Hatcheries on the Milwaukee River

One place where sturgeon are being restored in a slightly different way than a traditional fish hatchery is the Milwaukee River. That river, which flows out to Lake Michigan, hasn't had sturgeon naturally since the late 1800s, when overfishing, pollution, and dams caused them to die out.

In 2005, Wisconsin and Michigan began experimenting with mobile hatchery labs. In these "streamside" trailers, people raise the sturgeon in the same way as the hatcheries do. However, because the mobile labs sit on the banks of rivers, the water from the river flows through the tanks into the trailers where the sturgeon grow. Basically, the streamside hatcheries are like covered, climate-controlled detours in the river.

Naturalist Mary Holleback runs the streamside hatchery at Riveredge Nature Center along with the DNR. According to her, in the spring the hatchery gets sturgeon eggs from the Wolf River, right below the Shawano dam. Then the scientists raise the fish from eggs until they're fingerling size.

Aaron Schiller, a DNR fisheries biologist based in Milwaukee, explains the idea behind the streamside hatcheries: "They hatch in the Milwaukee River water and are raised in that water until we release them [into the Milwaukee River] at Sturgeon Fest. They imprint on that river water, just like a salmon does." Schiller notes that it may be the smell of the river that the sturgeon imprints on, and that the first three months of a sturgeon's life are the most important in imprinting onto a specific body of water.

It's too soon to say if this strategy will be successful (remember, it takes decades for sturgeon to mature enough to reproduce), but all

Mary Holleback oversees the Milwaukee River streamside hatchery at Riveredge Nature Center.

signs look promising for bringing sturgeon back to the Milwaukee River. As Schiller says, "There hasn't been any indication of spawning yet because we're just getting to the time period where it's going to happen."

Yet again, time is an important factor with these fish. Think about how long it takes for humans to set the pieces in place and wait for the fish to recover. Destroying a species can happen so quietly and so quickly, but it takes generations of people and a whole lot of patience, cooperation, and noisy effort to bring them back.

Caviar Scandal

When Lake Winnebago spearers nab a female sturgeon that is full of eggs, they have three choices for what to do with her eggs. They can keep them to make caviar, give the eggs to someone as a gift, or donate them to the DNR for research.

What they can't do is sell or trade the eggs or caviar. It's against the law. This law protects the sturgeon by prohibiting people from poaching fish to make money from the eggs.

In 2021, some employees of the Wisconsin DNR itself were accused of breaking that law. According to the law, once the DNR has finished using donated eggs for research, employees can either give the eggs back to the person who donated them or destroy them. Instead of doing either of those things, some DNR employees had the eggs made into caviar. They gave some of the eggs to the people who processed the caviar as a "payment." The DNR employees ate the caviar at meetings and other gatherings. It was a practice that had been going on for years. After all, if the eggs were just going to be destroyed, what was the harm in processing them into caviar? No one sold the eggs or made money from them. But the law was clear, and the DNR employees had broken it. One employee resigned and was arrested for lying about the practice. Three community members were charged with crimes for taking sturgeon eggs as payment for processing them into caviar.

Was what the people did wrong? Was the law wrong? What do you think?

CHAPTER 8

STURGEON CONSERVATION TODAY

Roxanne Komanec threads not one, not two, but three squirming nightcrawlers onto the large hook at the end of her catfish pole—which she's using today to catch sturgeon. She jokes that the mass of worms "should look like a steak or a meatball." She's in a boat on Lake Holcombe in northwestern Wisconsin, where there is a four-week hook-and-line lake sturgeon fishing season in September. (Fishing for sturgeon by hook-and-line is allowed in certain areas of the state at specific times of year. The Wisconsin DNR sets these regulations.)

She threads a sinker on next. It will pull the nightcrawler meatball to the bottom of the lake, where hopefully it'll catch the attention of a sturgeon. "The best thing for me is when you get that fish," Komanec explains. "We lift up all the anchors because they're smart fish. They'll come toward the boat and wrap around the anchors. Without the anchors down, they will pull the boat all over the place. They are so strong. It is unbelievable."

Komanec's fascination with sturgeon started when she was two years old, when her dad hooked his first sturgeon on the Flambeau River. Decades later, when her father started losing his sight, they traded roles, and she'd take him out sturgeon fishing. Eventually, friends, daughters, and granddaughters joined her. "What started as a father/daughter tradition has now evolved to an epic group of ladies," she says. They're known on the lake as the Lady Legends of Lake Holcombe.

In 2023, the Lady Legends had a record year. They caught eighty-nine sturgeon during the twenty-eight-day season. They've released every sturgeon they've caught during the twenty-two years they've been fishing together—except one. "To find out what it tastes like," Komanec explains. "We got it smoked. And it tasted like ham."

Three generations of Lady Legends of Lake Holcombe—Roxie Komanec, Amanda Schindler, and Avery Schindler—enjoy the hook-and-line sturgeon fishing season near Chippewa Falls, Wisconsin.

This love of the fish that Komanec and the Lady Legends have, the camaraderie, the traditions—they're all part of what Margaret Stadig, the Lake Winnebago system fisheries biologist, sees as essential to the lake sturgeon's long-term survival in Wisconsin. "I want more people to experience them," Stadig says. "I do hope that in the future as populations continue to rebound, more anglers might actually have a chance to go out and interact with these fish and build that respect." She's not the only one who thinks so, either.

The Keys to Lake Sturgeons' Survival

In 2019, a group of sturgeon stakeholders—biologists, enthusiasts, Sturgeon for Tomorrow members, Tribal leaders, and more—worked together for months to update a document called "Wisconsin's Sturgeon Management Plan." In it, they identify six issues critical to sturgeon survival in Wisconsin.

One of the first is to continue learning about sturgeon's life history and to build on what past Wisconsin fisheries biologists have discovered about sturgeon. "We're still missing so much of their life," Stadig says.

While the DNR has been tracking sturgeon with PIT tags for decades, historically PIT tags only provided information when and if a sturgeon was recaptured. Over the last few years, scientists have installed PIT tag antennas in waterways so that anytime a tagged sturgeon swims past, it pings the antenna. Aaron Schiller, the Milwaukee County fisheries biologist, explains how it works on the Milwaukee River: "Theoretically, if a fish gets all the way up to Mequon-Thiensville, I'll be able to detect them on all three antennas and plot their progress."

An even better (and pricier) tracking method are acoustic tags. These tags send out a signal constantly, so the biologists can tune in and track sturgeons' movements. It works kind of like GPS on a cell phone.

Scientists insert a PIT tag into a captured sturgeon so they can track it.

"But," Stadig says, "those tags have an expiration date." Their batteries last only about ten years. "We are still getting ten years of data out of it, which is incredible, but when you think about lake sturgeon, that's still only 10 percent of their life. So what we're trying to do is recapture some of these fish to put a new tag in so we can get twenty years of data."

She laughs. "I don't know about you, but . . . I'm not the same person I was ten years ago or twenty years ago. And so, I'm really intrigued to see whether or not we see any of these behavioral changes."

Also identified in the management plan is the need to focus on sturgeon habitat. "**Rehabilitation** is not just reintroducing them, [it's] reintroducing them in a fashion that the end goal is natural reproduction," Schiller says. The goal is a self-sustaining population so that humans no longer have to stock lakes with fish.

To become self-sustaining, sturgeon need both good water quality and space to migrate. "One of the biggest threats is habitat destruction and habitat fragmentation," Stadig says. "Habitat fragmentation is basically when you take an entire ecosystem and then you put a barrier in. Lake sturgeon require big, gigantic lakes . . . as well as hundreds of miles of river."

Habitat fragmentation cuts off access to part of the ecosystem for many species—and often it's a part they must live in to survive. That's historically what happened to sturgeon in Wisconsin when people started building dams throughout the state, keeping the fish away from the rivers where they needed to spawn.

Stadig continues: "So, as we move forward, it's important to make sure we don't destroy the lakes and habitat that is currently there. We don't want to fragment any more of their habitat." It is easier and less expensive to protect existing habitat than to try to create or restore new habitat later.

In some areas of the state, people have begun removing barriers that the fish have faced. The Milwaukee River is an example. In recent years, dams on the river have been removed or modified to allow fish passage. These fishways around the dams both connect the fragments of the sturgeon's historical habitat and provide a scenic area for people to observe the fish.

But it's in the far northeastern part of the state where you'll find the most dramatic fish passage—a fish *elevator* on the Menominee River. In the spring and fall, fish swim up to the dam and enter the underwater elevator. The door shuts behind them and the elevator shuttles them up to a fish sorting area at the top of the dam. Here, people scan, tag, and measure the sturgeon and record data about them. Spawning sturgeon are placed in special trailers and trucked around a second dam upriver. Other fish species that happen to be caught in the elevator are simply released back downstream.

It's a unique but labor-intensive solution that came to be through the combined efforts of Wisconsin's and Michigan's state governments

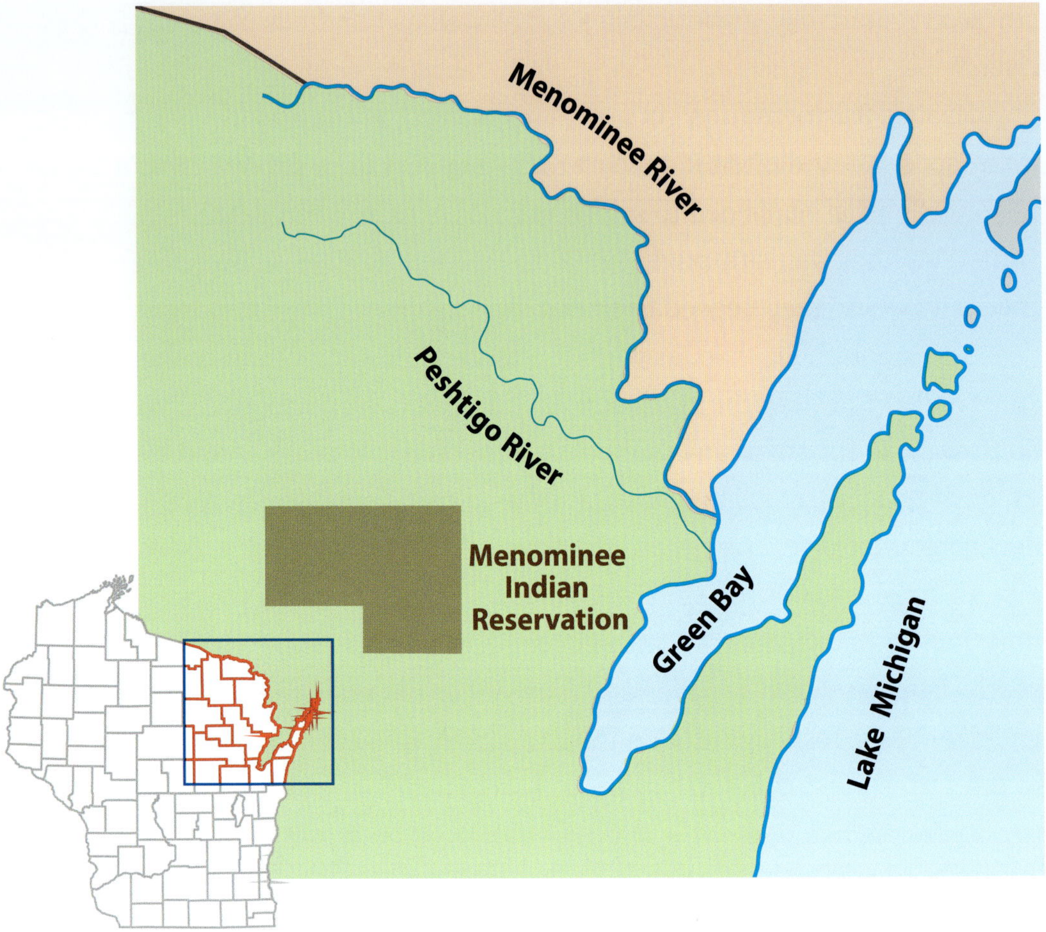

and the US Fish and Wildlife Service. And it's the kind of creative thinking and cooperation that species like lake sturgeon need.

Improving the existing habitat is a constant project, too. Back in the 1960s and 1970s, people along the Wolf and Fox Rivers started adding riprap (loose piles of stones) to their shorelines to prevent erosion. Later they discovered the riprap attracted spawning sturgeon. It provided new habitat for them to lay their eggs.

Ozaukee County, through which the Milwaukee River flows, is also working to improve the habitat for future sturgeon. The county recently completed sonar scanning of its part of the river to make sure there is enough of that rocky base so that everything is in place when sturgeon start spawning there.

But for every step forward Wisconsin takes in providing better habitat for sturgeon, a new challenge crops up.

To List or Not to List?

Worldwide, the International Union for the Conservation of Nature (IUCN) has listed lake sturgeon as "endangered" due to their declining populations for the last three generations. In 2023, the US Fish and Wildlife Service considered listing them as an endangered or threatened species in the United States. It was a controversial period for sturgeon lovers in Wisconsin. Any change in federal status could affect the state's fishing and spearing regulations and traditions.

In December 2023, a group of nine Wisconsin congressional leaders came together to address the USFWS. They wrote, "Nowhere in the world will you find such a unique cultural connection and staunch [loyal] dedication to the preservation of sturgeon population levels than in Wisconsin. We are concerned that a potential listing of the species under the Endangered Species Act could curtail [stop] this successful, science-based management model as well as threaten a cherished and unique Wisconsin tradition."

Mike Gallagher, a US congressional representative for northeastern Wisconsin, followed up the statement with the introduction of a bill that would exempt Wisconsin from any endangered listing for lake sturgeon.

While the sturgeon population is healthy in Lake Winnebago, that isn't true for lake sturgeon populations in other parts of Wisconsin and in other states. It can be difficult to meet the needs and wants of people (and fish) everywhere in the country.

"It's not just cultural or economic connection[s] that have made our community the sturgeon capital of the world," Gallagher said. "We also have the world's most meticulous [careful] management plan. It's one that combines public and private efforts from biologists at the DNR to fishermen and local fishing clubs up and down Lake Winnebago."

While it's true that the Lake Winnebago population of sturgeon is healthy and sustainable, conservationists argued that other sturgeon populations in the state were in danger. Lake sturgeon numbers in the US have been greatly reduced over the last hundred years. A statewide exemption could further endanger those fish, conservationists cautioned.

After a yearlong investigation and debate, USFWS announced in 2024 that lake sturgeon do not require listing under the Endangered Species Act. Because of this, the state of Wisconsin will continue to make its own rules and regulations concerning sturgeon. But even though sturgeon were not declared endangered, the ruling has drawn attention to other looming threats to sturgeon and their habitat across their entire range—threats like climate change.

Warmer temperatures and erratic weather patterns mean changing water temperatures for the fish. "Sturgeon do have a temperature tolerance," Stadig says. "If it gets too warm, or it gets too cold, their likelihood of surviving decreases. Maybe not so much for the adults, but for the juveniles and their eggs. So, over time, you'd see populations decline."

Fish diseases can also increase when water temperatures change. At the streamside hatchery at Riveredge in Saukville in 2023, only about half of the fingerlings survived. "I'm predicting that those fish were exposed to **pathogens** that their immune systems were not prepared for because it was not at the timing that they normally hit warm water," Milwaukee County biologist Schiller says.

As we've seen in the past with sturgeon, a setback in one stage of development leaves a gap in the whole species' population for their lifespan.

A hundred years. That's how long it will take to carefully and continually monitor the fish to make sure we don't repeat our past mistakes with sturgeon.

How You Can Help

Remember the "Wisconsin Sturgeon Management Plan" from 2019? One of the most important goals mentioned in that document is "public outreach and engagement." Simply put, the more people know about sturgeon, the more people will appreciate them. The more people appreciate them, the more they will make choices and sacrifices to help them survive. But "public outreach and engagement" doesn't just mean "other people." It also means *you*.

Sturgeon in Wisconsin need *your* help. Think of how much you know now about these amazing living dinosaurs! You can go forth and tell your friends and family and neighbors about them and why they should care about them.

That's outreach. How about engagement?

Stop by a hatchery in Wild Rose, Genoa, or Saukville to see sturgeon being raised. Check out the brine shrimp and krill they eat. Feel the sharpness of their new scutes.

Venture out in February to a sturgeon spearing registration station on Lake Winnebago. Listen to the spearers tell stories of their catches. Marvel at the enormous fish being hefted onto the scales.

In April, check the DNR's sturgeon webpages to learn when the sturgeon will start spawning on the Wolf and Fox Rivers. Join the crowds on the riverbanks. Watch the biologists net and tag the fish. Get splashed by the sturgeon surging upriver.

Can you volunteer at your local fish hatchery, like this Youth Conservation Corps volunteer at the Genoa National Fish Hatchery in Genoa, Wisconsin?

Swing by Keshena in April, too. Watch the Menominee fish dance celebrating the sturgeon at the nation's spring powwow. Tour the Menominee Cultural Museum to find out more about the clans. Taste some sturgeon.

In September, check the DNR website for a hook-and-line sturgeon fishing lake near you. Load a hook with nightcrawlers and see what you pull up out of the water. If you're lucky, feel a tug on your line.

Also in September, attend Sturgeon Fest in Milwaukee. Release your own sturgeon fingerling into the Milwaukee River. Register your sturgeon so you can be notified when (or if) its PIT tag is rescanned.

Don Wojahn sends a fingerling sturgeon into its new home in the Milwaukee River during Sturgeon Fest. A PIT tag and a special website allows him to check in on the fish as it grows.

Doing some of these things might inspire you to consider exploring a career in fisheries or aquatic science someday.

Or how about volunteering? You can do that right now. Kyrie Rietz, age eleven in 2025, and her parents, Joshua and Jennifer Rietz, volunteer with the Lake Winnebago Sturgeon Guard. They hand out information about sturgeon and the Sturgeon Guard at local events.

For more than thirty years, the DNR ran the Sturgeon Guard. The agency trained volunteers to watch over the rivers during the spring spawning and report any incidents of poaching or harassment. The Sturgeon Guard ended during the COVID-19 pandemic in 2020. But in 2023, Don Mielke, a Sturgeon for Tomorrow member and fifth-generation spearer, decided to revive the Sturgeon Guard as an all-volunteer organization. Not only do volunteers report suspicious activity, but they also serve as information sources along the rivers, educating people about what they're watching. Kyrie has even gotten the chance to assist the biologists by holding the sturgeon and helping with the recording.

It's not just the Lake Winnebago area where people have organized to help sturgeon. The Sturgeon Protectors in the Milwaukee area teach people about the sturgeon they will find in the Milwaukee River. Local chapters of River Alliance, an organization dedicated to protecting Wisconsin's waters, have helped revive the sturgeon population in the St. Louis River and assisted with the fish elevator project on the Menominee River. There are smaller groups that help, too, from nature centers to conservancy organizations.

Sturgeon were once found in almost all Wisconsin's rivers and big lakes. Who's to say you can't help bring them back to your area?

As Mary Holleback, the naturalist and citizen science manager at Riveredge Nature Center, explains, "Kids are going to be the ones that will end up being the caretakers for the fish." We need to get the word out that these are "different, interesting, and important" fish to protect.

Across the state, Roxanne Komanec, a Lady Legend of Lake Holcombe, would agree. "The sturgeon is so unique," she says. "There is no more beautiful fish in the world."

So go outside and do what you can to help everyone in Wisconsin enjoy them!

Margaret Stadig, Wisconsin's "Sturgeon General"

Wisconsin's head sturgeon biologist, Margaret Stadig, with a lake sturgeon (and bobber earrings) on Lake Holcombe

In 2022, a woman took on the job of Wisconsin's head sturgeon biologist for the first time ever. In fact, at the time Margaret Stadig was one of just a few female fisheries biologists in the Wisconsin DNR and in the fisheries field in general. As Stadig explains it, "I'm not here to be the representative for all women. I'm here to do a job that I trained for. But at the same time, I am here. And I am the first."

Stadig bubbles with expertise and enthusiasm about sturgeon. But she's also thoughtful about how being the first woman in this role might benefit not only other women and other historically excluded groups entering the field but the fish, too. Stadig explains:

> I want to see more females in this field. We bring a different take. We bring a different energy. We bring a different perspective. Not that the old perspective is bad. It's just different. And we need *different* more now in fisheries than ever as funding shrinks and we're being asked to do more with less. We have to be able to think of new, creative ways to do things while still understanding why the old ways work and acknowledging them. It's about using what has worked in the past but guiding it toward a potentially new and interesting future. And that's not just with women. That's with all minorities. Because everyone interacts with the environment differently, and we can't just continue to cater to one group. We have to consider all of these other individuals who are interacting with nature and these fish to make sure that everyone feels involved and everyone feels the same kind of passion.

That's what will help save the sturgeon—making sure their survival is important to everyone in Wisconsin and including everyone's voices in the conservation. And Stadig is helping to move that forward.

Acknowledgments

Many thanks are due to the following individuals:

- Kathleen Schmitt Kline, Ron Bruch, and Fred Binkowski, authors of *People of the Sturgeon* (Wisconsin Historical Society Press, 2009), and photographer Bob Rashid. Their book served as a crucial source for this one, as did the many resources that the authors collected and compiled.

- Senior Special Librarian and Education Coordinator Anne Moser and Fisheries Specialist Titus Seilheimer, both of Wisconsin Sea Grant, who conducted a thorough and insightful review of this book in manuscript form, correcting and clarifying many of the scientific concepts and historical details. Extra thanks to Anne Moser, who came to the WHS Press with the idea for a sturgeon conservation book for young readers, and for her participation and enthusiasm during every stage of the book's development.

Places in Wisconsin to See Sturgeon

Genoa National Fish Hatchery, Genoa,
www.fws.gov/fish-hatchery/genoa

Menominee Indian Tribe Cultural Museum, Keshena,
www.menominee-nsn.gov/CulturePages/CulturalMuseum.aspx

Riveredge Nature Center, Saukville,
www.riveredgenaturecenter.org

Sturgeon Fest, Milwaukee,
sturgeonfest.org

Sturgeon hook-and-line fishing season (September), specific rivers/lakes,
dnr.wisconsin.gov/topic/Fishing/sturgeon/SturgeonInland Fishery

Sturgeon spawning viewing sites (during spawning, usually mid-April/May), Shawano, Shiocton, and New London,
dnr.wisconsin.gov/topic/Fishing/sturgeon/SturgeonSpawning.html

Sturgeon spearing season (February), Lake Winnebago system,
dnr.wisconsin.gov/topic/fishing/sturgeon/WinnSysSturgeon Spear

Wild Rose State Fish Hatchery, Wild Rose,
dnr.wisconsin.gov/topic/Fishing/hatcheries/wildrose.html

Volunteer to Help Sturgeon and Their Habitat

River Alliance of Wisconsin (headquartered in Madison, with branches around the state), **wisconsinrivers.org**

River Keepers (Milwaukee), **milwaukeeriverkeeper.org**

Riveredge Nature Center (Saukville), **www.riveredgenaturecenter.org**

St. Louis River Alliance (Duluth/Superior), **www.stlouisriver.org**

Sturgeon Guard (Lake Winnebago system), **dnr.wisconsin.gov/topic/Fishing/sturgeon/SturgeonGuard.html**

Wild Rivers Conservancy (Osceola), **wildriversconservancy.org**

Check with your local environmental organizations or organize something yourself! Sturgeon used to be common in most Wisconsin rivers and lakes.

Can you help rehabilitate them in your area?

Who's Who

Indivduals

Abraham, Robert "Jake": Fished for sturgeon when it was banned between 1915 and 1932

Addis, Jim: DNR fisheries chief in the 1970s who agreed to try raising sturgeon

Ballard, William: Dartmouth College professor who had been to Russia and knew the "recipe" to raise sturgeon from eggs

Binkowski, Fred: UW–Milwaukee scientist who figured out what young sturgeon would eat in captivity

Bruch, Ron: Oshkosh fish manager in 1990; responsible for many sturgeon research findings

Casper, Bill: Founder of the Sturgeon for Tomorrow group in the 1970s

Cox, Douglas: Director of land management at the Menominee Indian Tribe of Wisconsin; past and present advocate for fish passage on the Wolf River

Czeskleba, Don: Wild Rose DNR hatchery supervisor in the 1970s; present for first successful hatching of eggs

Daley, Russ: DNR biologist in the 1970s; also the Clay Man who helped with the first attempts to raise lake sturgeon in a hatchery

Devine, Barney: Chief warden of the state conservation department during the early 1900s

Folz, Dan: Fish manager at Oshkosh in the 1970s; also the Egg Man who helped with the first attempts to raise sturgeon in a hatchery

Gorchals, Ed: Arrested for catching sturgeon illegally during the time it was banned from 1915 to 1932

Graass, Frank N.: A lobbyist from the Izaak Walton League conservation group in the 1930s

Grignon, Dave: Tribal historic preservation officer for the Menominee Nation

Holleback, Mary: Naturalist who manages the streamside hatchery at Riveredge Nature Center

Kelleter, Paul: Wisconsin Conservation Commission director in the 1930s

Komanec, Roxanne: Avid sturgeon hook-and-line fisher; member of the Lady Legends of Lake Holcombe

Lang, Reverend Walter: Owned a private trout farm in New London that was the site of the first attempt to raise sturgeon in 1978

Leppla, Howard: Conservation Congress representative from Calumet County in the 1930s; argued for keeping a spearing season

Mauthe, William: Chairman of the Wisconsin Conservation Commission from 1927 to 1933

Mielke, Don: Fifth-generation sturgeon spearer and organizer of the 2023 rebooted Sturgeon Guard

Milner, James: US Commission of Fish and Fisheries Great Lakes deputy commissioner in the 1870s; supported the sturgeon industry along the lakes

Ogden, Lee: Arrested for smuggling sturgeon to Chicago in 1916

Pamonicutt, Pete: Menominee Tribal court judge in the 1930s who asked for sturgeon to be stocked above the dams

Priegel, Gordon: Fisheries biologist who developed the 5 percent harvest limit in 1967

Primising, Mike: DNR biologist in the 1970s; also the Pan Man who helped with the first attempts to raise sturgeon in a hatchery

Reiter, Don: Fish and wildlife manager for the Menominee Nation who was involved in the plan to return sturgeon to the reservation and subsequent sturgeon monitoring efforts

Rietz, Jennifer, Joshua, and Kyrie: Sturgeon Guard volunteers

Schiller, Aaron: Senior DNR fisheries biologist based in Milwaukee

Schneberger, Edward: Wisconsin's first fisheries biologist; started the collection of sturgeon data in the 1940s

Smith, Hugh M.: US Commission of Fish and Fisheries Great Lakes deputy commissioner in 1914; warned of the demise of lake sturgeon

Stadig, Margaret: Lake Winnebago system fisheries biologist for sturgeon

Stauflien, Sam: Wild Rose Fish Hatchery fish culture biologist

Groups

Conservation Commission: State agency established in 1915 to oversee all fish, wildlife, and natural resources in Wisconsin; the Fisheries Commission become part of this organization in the 1930s and in the 1960s it turned into the Department of Natural Resources

Conservation Congress: Group of representatives for each Wisconsin county that provides citizen input to the state DNR (formerly Conservation Commission) on natural resource management

Department of Natural Resources (DNR): State government agency that manages fish, wildlife, and natural resources for Wisconsin, established from the state Conservation Commission reorganization in 1967

Fisheries Commission: First state government agency established to survey the fish of Wisconsin

International Union for the Conservation of Nature (IUCN): Global organization that determines the conservation status of species in the natural world (near threatened, endangered, extinct, and so on)

Menominee Indian Tribe of Wisconsin: Indigenous nation living in Wisconsin today and whose ancestors have lived in northeastern Wisconsin and western Michigan for thousands of years

Sturgeon for Tomorrow: Organization made up of sturgeon spearers and other citizens that began in the Lake Winnebago area; driving force behind getting lake sturgeon raised in hatcheries

Sturgeon Guard: Volunteers who watch over the sturgeon during spawning season and report any poaching to authorities

Sturgeon Protectors: Volunteer organization in the Milwaukee area that educates others about sturgeon rehabilitation in the Milwaukee River

US Fish and Wildlife Service (USFWS): Federal government agency that oversees management of the fish and wildlife of the United States

Timeline of Events

Before the Common Era (BCE)

About 100 million years ago Lake sturgeon first appear

14,000 years ago Glaciers melt, creating new lakes and rivers

10,000 years ago Menominee people's history begins

1800

1860

1860s–1890s Sturgeon overharvested for caviar

1870

1871 Congress establishes US Fish Commission (later becomes US Fish and Wildlife Service)

1874 Wisconsin's first Fisheries Commission surveys state's fish

1876 First Wisconsin state fish hatchery established

1880

1890

1892 Shawano dam built on Wolf River

1900

1950s Conservation Commission continues to assess and research sturgeon

1959 Last sturgeon speared on Menominee Reservation

1960

1967 Wisconsin Conservation Commission becomes Wisconsin Department of Natural Resources

1970

1973 Endangered Species Act becomes law

1977 Sturgeon for Tomorrow founded

1979 First successful attempt to raise sturgeon in a hatchery in Wisconsin

1980

1983 Sturgeon stocked for first time in St. Louis River

1988 Public volunteers begin participating in DNR-run Sturgeon Guard program

1990

1993 DNR delivers sturgeon to Menominee Reservation; first Menominee sturgeon celebration in 100 years held

1995 Reintroduction plan for sturgeon on Menominee Reservation established

2000

2003 Sturgeon rehabilitation project begins on Milwaukee River

Common Era (CE)

1810 · **1820** · **1830** · **1840** · **1850**

- **1809** First dam built in Wisconsin
- **1836** Wisconsin Territory formed
- **1839** Law passes requiring dams to include fish passage
- **1840** Milldam Act passes, negating 1839 fish passage law
- **1848** Wisconsin becomes a state
- **1854** Menominee Reservation created

1910 · **1920** · **1930** · **1940** · **1950**

- Wisconsin Fisheries Commission becomes part of Conservation Commission
- **1915** Sturgeon harvesting banned in Wisconsin
- **1926** Balsam Row dam built on Wolf River
- **1932** Regulated spearing season opens on Lake Winnebago
- **1933** Conservation Commission takes over creation of fishing regulations from state legislature
- **1935** Sturgeon stocking on Menominee Reservation begins
- **1938** Spearing community around Lake Winnebago asks Conservation Commission to base fishing regulations on data
- **1941** Conservation Commission begins collecting fish data

2010 · **2020**

- **2006** Streamside hatcheries begin in Wisconsin and Michigan
- **2006** Menominee Nation wins lawsuit for creation of fish passage at Balsam Row dam
- **2012** Sturgeon spawn at Keshena Falls for the first time in over 100 years
- **2017** Menominee River fish elevator completed
- **2019** DNR-run Sturgeon Guard ends
- **2019** Wisconsin's Sturgeon Management Plan updated
- **2022** IUCN lists lake sturgeon as "endangered"
- **2023** Citizen volunteers reestablish Sturgeon Guard
- **2023–2024** USFWS declines to list lake sturgeon as endangered

Glossary

barbels: "Whiskers" on a sturgeon's face that help it sense food

cartilage: Firm but flexible tissue that sturgeons have instead of a skeleton

caviar: Processed food made of salted fish eggs

census: A counting of a certain population of animals, people, or plants

commission: Group of people directed to perform some duty

conservation: Careful preservation and protection of nature

dorsal fin: A fin along the middle of a fish's or marine mammal's back

ecosystem: System made up of all the living things that interact with the surrounding environment

euthanize: To humanely kill an animal without causing pain

extirpated: When a species is no longer found in an area where it used to live but still exists elsewhere

fertilize: When a sperm and an egg join to create new life

fingerling: Young sturgeon about the size of a finger and about six months old

fishway or **fish passage:** Construction that allows fish to swim past a dam and reach the other side

genetic diversity: Range of traits within a species; a larger population means a higher genetic diversity and allows species to better adapt to a changing environment than a smaller population might

hatcheries: Places where people control the hatching of fish or poultry eggs for commercial or conservation purposes

indicator species: Species of plant or animal that reflects environmental change in its habitat

Indigenous: Referring to people who have historically lived in a particular place without immigration

krill: Tiny, shrimp-like creatures that float in the ocean in large quantities; they are fed to young sturgeon in fish hatcheries

larvae: Newly hatched insects or fish

legislature: Group of people elected to make laws

mature: Old enough to reproduce

milt: Fluid inside a male fish that contains sperm

notochord: Cartilage-encased, flexible rod that sturgeon have instead of a spinal cord

otolith: Part of the inner ear of sturgeon; used to determine how old a sturgeon is

pathogens: Germs that cause disease

PIT (passive integrated transponder) tag: Small device with a microchip placed inside an animal's body that allows people to track that animal's movements and behavior

poach: To fish or hunt when or where it is illegal to do so, or by an illegal method

rehabilitation: The act of restoring something back to health and good condition

representatives: Elected lawmakers

scutes: Protective bony plates arranged in five rows along a sturgeon's body

shanties: Small, moveable shacks that ice fishers and spearers use for shelter

spawn: When a female fish releases eggs or a male fish fertilizes the eggs; the eggs that are released

species: Category of living things that are made up of related individuals able to reproduce

stock: To release hatchery-raised fish into a body of water

strains: Groups of plants or animals that look alike but have characteristics that make them slightly different

sustainably: Relating to harvesting or using a resource in a way that ensures it is not damaged or used up

telemetry: In wildlife research, a way of recording and transmitting data on animal movement and behavior; telemetry can use acoustics (sound), radio signals, or global positioning system (GPS) signals

watershed: The area that drains into a river or a lake

yearling: Young sturgeon that is a year old and about eleven inches long

Selected Bibliography

Asterisks (*) indicate sources that young readers may find useful to learn more about sturgeon conservation.

Author Interviews

Cox, Douglas, telephone interview with the author, November 13, 2023.

Grignon, David, telephone interview with the author, November 20, 2023.

Holleback, Mary, telephone interview with the author, October 13, 2023.

Komanec, Roxanne, telephone interview with the author, October 10, 2023.

Rietz, Jennifer, Joshua Rietz, and Kyrie Rietz, telephone interview with the author, October 14, 2023.

Schiller, Aaron, telephone interview with the author, October 12, 2023.

Stadig, Margaret, telephone interview with the author, October 5, 2023.

Stauflien, Sam, interview with the author, Wild Rose, WI, July 21, 2023.

Books

*Bosanko, David. *Fish of Wisconsin Field Guide.* Cambridge: Adventure Publications, 2007.

*Kline, Kathleen Schmitt, et al. *People of the Sturgeon: Wisconsin's Love Affair with an Ancient Fish.* Madison: Wisconsin Historical Society Press, 2009.

Loew, Patty. *Indian Nations of Wisconsin: Histories of Endurance and Renewal*. 2nd ed. Madison: Wisconsin Historical Society Press, 2013.

Oral History Interview Recordings

All of the following are part of "People of the Sturgeon Interview Collection," Wisconsin Historical Society and Oshkosh Public Museum, **digital.library.wisc.edu/1711.dl/SturgeonPrj**:

- Abraham, Robert, interview by Dick Ristow, March 20, 2006.
- Binkowski, Frederick P., interviews by Kathleen Schmitt Kline, July 10, 2008 and August 14, 2008.
- Bruch, Ronald M., interview by Paul Muche, September 24, 2007.
- Bruch, Ronald M., interview by Kathleen Schmitt Kline, August 15, 2008.
- Casper, Bill, interview by Paul Muche, April 11, 2007.
- Corbett, Ken, interview by Ronald M. Bruch and Kathleen Schmitt Kline, May 27, 2008.
- Diemel Jr., Haze, interview by Ronald M. Bruch and Kathleen Schmitt Kline, May 27, 2008.
- Folz, Dan, and Mike Primising, interview by Bob Marin, April 16, 2007.
- Gorchals, Ed, interview by Dick Koerner, October 12, 2007.
- Priegel, Gordon, interview by Ronald M. Bruch, October 25, 2007.

Other

Milwaukee Journal Sentinel* frequently publishes news and feature stories on lake sturgeon in Wisconsin: **www.jsonline.com/search/?q=%22lake+sturgeon%22.

2024 Winnebago System Sturgeon Spearing Regulations. Publication no. FH-305-2023. Wisconsin Department of Natural Resources, 2024. **widnr.widen.net/s/hrzbtcnqpg/2024sturgeonspearingregulations**

Wisconsin's Sturgeon Management Plan. Bureau of Fisheries Management, Wisconsin Department of Natural Resources. Report no. 90, October 2019. **widnr.widen.net/s/qcp5gh68km/sturgeon_sturgeonmanagementplan2019**

Videos

Langston, Nancy. "Welcoming Back Namao: Indigenous Communities and Restoration of the Great Lakes Sturgeon." Lecture presented at Brandeis University, Waltham, MA, March 13, 2019. **www.brandeis.edu/mandel-center-humanities/lectures-videos.html#sturgeon**.

*"Nama'o The Ancient Story Teller." YouTube. Posted by Reynaldo A. Morales, December 19, 2017. **youtu.be/c5vII8Qzi3c?feature=shared**.

Roberts, Craig. "Lunch and Learn: Lake Sturgeon of the St. Croix and Namekagon Rivers." Lecture presented at Wild Rivers Conservancy, Osceola, WI, February 15, 2023. YouTube. Posted by Wild Rivers Conservancy, February 15, 2023. **youtu.be/s96_PHD9jrM?feature=shared**.

*Wolf River Cam: Underwater video streaming during spawning season. **www.wolfrivercam.com**.

Websites

Menominee Indian Tribe of Wisconsin. "Menominee Indian Tribe Cultural Museum." **menominee-nsn.gov/CulturePages/CulturalMuseum.aspx**.

Michigan Department of Natural Resources. "Lake Sturgeon Management." **www.michigan.gov/dnr/managing-resources/fisheries/sturgeon**.

*Michigan State University Board of Trustees. "Lake Sturgeon and Coupled Great Lakes-Tributary Ecosystems." **www.glsturgeon.com**.

*National Wildlife Federation. "Lake Sturgeon." **www.nwf.org/Educational-Resources/Wildlife-Guide/Fish/Lake-Sturgeon**.

On Wisconsin Outdoors. "Welcoming Back Namao." **www.onwisconsinoutdoors.com/Content/files/inlandFishing/SturgeonReturn.pdf**.

*Riveredge Nature Center. "Returning the Sturgeon." **riveredgenc.maps.arcgis.com/apps/Cascade/index.html?appid=b415d020b7ea4461957023e765ba963b**.

*Sturgeon for Tomorrow, Black Lake, Michigan Chapter. "Sturgeon in the Classroom." **www.sturgeonfortomorrow.org/sturgeon-in-the-classroom.php**.

*University of Wisconsin Sea Grant. "Explore Lake Sturgeon." **www.seagrant.wisc.edu/our-work/focus-areas/education/great-lakes-and-marine-education-resources/explore-lake-sturgeon**.

*University of Wisconsin Sea Grant. "People of the Sturgeon Story Map." **maps.aqua.wisc.edu/storymaps/peopleofthesturgeon**.

*Wisconsin Department of Natural Resources. "Lake Sturgeon." **dnr.wisconsin.gov/topic/Fishing/sturgeon/LakeSturgeon.html**.

Source Notes

Chapter 1

All quotes from Margaret Stadig from the author's interview on October 5, 2023.

Chapter 2

it's estimated that eleven million sturgeon swam: Elizabeth M. Hay-Chmielewski and Gary E. Whelan, eds., *Lake Sturgeon Rehabilitation Strategy* (Ann Arbor: Michigan Department of Natural Resources, 1997), 11.

"During these few days of their run": Newcomb Spoor, "Fishing for Sturgeon," *Wisconsin Conservationist*, May 1919, 6.

"The story of the sturgeons is one of the most distressing": Hugh M. Smith, "The Passing of the Sturgeon," in *The American Natural History: A Foundation of Useful Knowledge of the Higher Animals of North America*, ed. William Temple Hornaday (New York: C. Scribner's Sons, 1914), 300.

"out of a shameful waste": James Milner, "Fisheries of the Great Lakes," in Report of the Commissioner for 1872 and 1873, comp. United States Fish Commission (US Government Printing Office, 1874), 73.

Sturgeon production went from 7.84 million pounds: Margaret Beattie Bogue, *Fishing the Great Lakes* (Madison: University of Wisconsin Press, 2000), 160.

"the most reckless, senseless fishing imaginable": Smith, "The Passing of the Sturgeon," 300.

"any artificial barrier": Dam Design and Construction, Wis. Stat. § 333.03(3) (Legislative Reference Bureau, April 2005, no. 592), 61, docs.

Wisconsin had almost four thousand dams: "Dams Overview," Wisconsin Department of Natural Resources website.

"They will pass decent spawning habitat": Aaron Schiller, interview with the author, October 2023.

Chapter 3

"The evidence of the decline of the fisheries": "Fifth Annual Report of the Commissioners of Fisheries of the State of Wisconsin, for the Year Ending December 31, 1878," in Governor's Message and Accompanying Documents, volume II (Madison, WI: David Atwood, 1879), 24.

"well organized and noisy group[s]": Barney Devine to W. C. Wise, January 15, 1935, Wisconsin Conservation Department subject files, 1917–1979, Series 271, Box 753, Folder 1, Wisconsin Historical Society Archives, Madison, WI.

"A big fat sturgeon was seen": *Oshkosh Daily Northwestern*, May 11, 1916.

"Illegal fishing on the Wolf": William Mauthe, "Lake Winnebago," William Mauthe papers, 1917–1942, Wis Mss AA, Box 1, Wisconsin Historical Society Archives, Madison, WI.

"They took my boat": Ed Gorchals interview, October 12, 2007.

"That's how we made extra money": Robert Abraham interview, March 20, 2006.

"doubtless hasten the complete destruction": Paul D. Kelleter to Gov. Philip F. La Follette, March 30, 1931, Philip Fox La Follette Papers 1876–1973, Box 7, Folder 3, Wisconsin Historical Society Archives, Madison, WI.

"During every session of the legislature": Frank N. Graass, interview by Dennis East, January 28 and 29, 1965, SC 972. Wisconsin Historical Society Archives, Madison, WI.

"From an enforcement angle": Excerpts of Transcripts of State Meeting of Conservation Committees Held at Madison, Wisconsin on July 12 and 13, 1938, Series 271, Box 753, Folder 2, Wisconsin Historical Society Archives, Madison, WI.

"We have the only sturgeon in the state": Excerpts of Transcripts of State Meeting of Conservation Committees Held at Madison, WI, July 12, 1938, Wisconsin Conservation Department subject files, 1917-1979, Series 271, Box 753, Folder 2, Wisconsin Historical Society Archives, Madison, WI.

"The fact is well known that fish and game": Reply to Resolution of Calumet County Sportsmen's Club from the Wisconsin Conservation Department, January 10, 1939, Wisconsin Conservation Department subject files, 1917-1979, Series 271, Box 753, Folder 3, Wisconsin Historical Society Archives, Madison, WI.

"It is our lake; we fish on it": Public Hearing Relative to Modification of Conservation Commission Order No. F-335 Dealing with Ice Fishing Regulations, January 17, 1939, Wisconsin Conservation Department subject files, 1917-1979, Series 271, Box 753, Folder 3, Wisconsin Historical Society Archives, Madison, WI.

"I asked Dr. Schneberger": Public Hearing Relative to Modification of Conservation Commission Order, Wisconsin Conservation Department subject files, 1917-1979, Series 271, Box 753, Folder 3, Wisconsin Historical Society Archives, Madison, WI.

Chapter 4

"I am rather firmly convinced": William J.K. Harkness, "Report on Artificial Propagation of Sturgeon," Wisconsin Conservation Department subject files, 1917–1979, Series 271, Box 669, Folder 7, Wisconsin Historical Society Archives, Madison, WI.

"I figured it was conservative": Gordon Priegel interview, October 25, 2007.

Chapter 5

"We had the life history and population": Dan Folz and Mike Primising interview, April 16, 2007.

"You think you know almost everything": Folz and Primising interview.

"What you're trying to do is raise sturgeon": Bill Casper interview, April 11, 2007.

"We entered [the meeting] with an open mind": Bill Casper interview.

"My God, man": Bill Casper interview.

"I know the recipe for this": Bill Casper interview.

"Guard this with your life": Bill Casper interview.

"Now we can relax": Folz and Primising interview.

"I see him get a scalpel": Folz and Primising interview.

"It must have something to do with the nutritional value": Frederick P. Binkowski interview, July 10, 2008.

"It's a testament to how far back": Ronald M. Bruch interview, August 15, 2008.

Chapter 6

"river at the place abundant": "History of the Namekagon," Namekagon River: Wisconsin's Moving National Park website, October 25, 2021.

All quotes from David Grignon from the author's interview on November 20, 2023.

"the poorest region in Wisconsin": Patricia K. Ourada, *The Menominee Indians: A History* (Norman: University of Oklahoma Press, 1979), 118.

"We used to have sturgeon on the reservation": Annual Indian Affairs Round Table, Oshkosh, WI, April 10, 1930, Wisconsin Conservation Department subject files, 1917–1979, Series 271, Box 419, Folder 1, Wisconsin Historical Society Archives, Madison, WI.

"We had federal, state, and Tribal biologists:" Don Reiter interview, August 25, 2008.

All quotes from Douglas Cox from the author's interview on November 13, 2023.

biologists can tell that more than 50 percent: Don Reiter, "Menominee Lake Sturgeon Management," lecture presented at 2013 Wisconsin Lakes Partnership Convention, Stevens Point, WI.

Chapter 7

All quotes from Sam Stauflien from the author's interview on July 21, 2023.

All quotes from Mary Holleback from the author's interview on October 13, 2023.

All quotes from Aaron Schiller from the author's interview in October 2023.

Chapter 8

All quotes from Roxanne Komanec from the author's interview on October 10, 2023.

All quotes from Margaret Stadig from the author's interview on October 5, 2023.

All quotes from Aaron Schiller from the author's interview in October 2023.

"Nowhere in the world": Paul A. Smith, "State Lawmakers Call for Sturgeon Exemption," *Milwaukee Journal Sentinel*, December 14, 2023.

"It's not just cultural or economic connection[s]": Joe Schulz, "Wisconsin's US Lawmakers Want to Exempt State Sturgeon From Possible Endangered Designation," Wisconsin Public Radio, January 23, 2024.

"public outreach and engagement": Bureau of Fisheries Management, *Wisconsin's Sturgeon Management Plan*, Wisconsin Department of Natural Resources, report no. 90, 38.

All quotes from Mary Holleback from the author's interview on October 13, 2023.

Index

Please note that page numbers in ***bold italics*** indicate illustrations.

K

L

M

Image Credits

All maps and diagrams by John Ferguson. **CHAPTER 1** Hauling netted sturgeon, measuring sturgeon, and releasing sturgeon, Rebecca Hogue Wojahn. Sturgeon in net, Brett Billings, US Fish and Wildlife Service. Sturgeon image from anatomy diagram, Wisconsin DNR. Spawning sturgeon, Bob Rashid. Juvenile lake sturgeon, full body and underside of head, Ryan Hagerty, US Fish and Wildlife Service. **CHAPTER 2** Mussel shell decoy, OPM #1998.22.220, Oshkosh Public Museum Archives, Oshkosh, Wisconsin. Bone harpoon, OPM #800-122, Oshkosh Public Museum Archives, Oshkosh, Wisconsin. 1888 Butchering float, Stefan Claessan, National Oceanic and Atmospheric Administration Photo Library. Men with sturgeon on line, OPM 2002.81, 1913, Oshkosh Public Museum Archives, Oshkosh, Wisconsin. Fox River dam, WHI Image ID 30438. McDonald fishway, Smithsonian Institution Archives, Acc. 11-007, Box 018, Image No. MNH-4208. **CHAPTER 3** Peshtigo sawmill and dam, WHI Image ID 78879. 1898 ice shanty, WHI Image ID 24511. Rolling gill net, Stefan Claessan, National Oceanic and Atmospheric Administration Photo Library. Snagging in 1870, WHI Image ID 78727. Man with pipe and sturgeon, Lester Public Library. Wardens with sturgeon, Reese Staber, UW Digital Collections, CC 4.0. Man with sturgeon on ice and woman with hanging sturgeon, Dean Tvedt, University of Wisconsin Digital Collections. Sturgeon decoys and spears, Bob Rashid. **CHAPTER 4** Postcard, otolith, and fin cross section, Wisconsin DNR. Men tagging sturgeon, WHI Image ID 132237. Metal tags, OPM Acc#2007.15, Oshkosh Public Museum Archives, Oshkosh,

Wisconsin. **Chapter 5** Bill Casper and shanty, WHI Image ID 123932. Spawning sturgeon, Bob Rashid. Collecting milt, Fred P. Binkowski, UWM School of Freshwater Sciences. Man at Wild Rose hatchery, WHI Image ID 132238. Eggs in jar, Fred P. Binkowski, UWM School of Freshwater Sciences. PIT tags and syringe, Bob Rashid. **Chapter 6** John Snow decoy, Wisconsin Historical Museum Object #1996.118.106A-B. Bear Trap Falls, WHI Image ID 61836. Menominee tobacco blessing, transferring sturgeon to pond, and Keshena Falls, Bob Rashid. **Chapter 7** Biologist cleaning tank, tank full of sturgeon, and brine shrimp, Ryan Hagerty, US Fish and Wildlife Service. Sturgeon life cycle diagram images: eggs, United States Geological Survey; free embryo, Kimberly Chojnacki, United States Geological Survey; larvae, Amy George, United States Geological Survey; juvenile, Ryan Hagerty, US Fish and Wildlife Service; adult, Wisconsin DNR. Mary Holleback at Riveredge, Michael Timm. **Chapter 8** Three women with sturgeon, Kelly Sayles. Inserting PIT tag and sturgeon fins above water, Bob Rashid. Young volunteer at tank, Ryan Hagerty, US Fish and Wildlife Service. Releasing fingerling, Rebecca Hogue Wojahn. Margaret Stadig, Eric Stadig.

About the Author

Rebecca Hogue Wojahn, also known as W.H. Beck, is the author of many books for children. Her books have been honored by state and regional awards, Junior Library Guild, the National Science Teachers Association, and Bank Street Best Books of the Year. A long-time Wisconsin resident and school librarian, she enjoys art, books, and exploring nature. Find out more at **www.whbeck.com**.